ACTA UNIVERSITATIS UPSALIENSIS
Studia Statistica Upsaliensia
3

Claes Andersson

On the Use of Two-Phase Sampling in Estimation of Parameters in Domains where Data Contain Misclassification and Measurement Errors

UPPSALA 1994

Doctoral Dissertation at Uppsala University 1994

Abstract

Andersson, C., On the Use of Two-Phase Sampling in Estimation of Parameters in Domains where Data Contain Misclassification and Measurement Errors. Acta Univ. Ups., *Studia Statistica Upsaliensia* 3. 120 pp. Uppsala. ISBN 91-554-3305-7.

In many surveys, the units in the population constitute a number of subgroups or domains, and comparisons are to be made between the domains regarding some response variables. When the variables used to classify the units into domains contain errors, misclassification will occur. Errors in the response variables are called measurement errors. In this thesis it is assumed that two devices are available to classify units and to get measurements of the response variables. The first, fallible device tends to misclassify data but may or may not measure the response variables correctly. The second, true device is a procedure that gives error-free classifications and measurements. To estimate parameters of interest an initial sample of n_a units is drawn, and the fallible device is used. A subsample of n units is drawn from the initial sample, and the true device is used in the subsample in order to get true classifications and measurements. The erroneous values of the initial sample are used as auxiliary information to improve the estimates from the data in the second-phase sample in a modified general regression estimator, *m*GREG. A Monte Carlo study is carried out to compare the *m*GREG estimator with other proposed estimators of the domain means. The *m*GREG estimator is approximately unbiased and its variance estimator behaves better than the variance estimators proposed for other approaches. It can also be used to estimate a general function of totals and/or under general sampling designs. In the thesis the multicharacter problem of allocating the second-phase sample to get efficient estimates in all domains of interest is also considered.

Claes Andersson, Uppsala University, Department of Statistics, Box 513, S-751 20 Uppsala, Sweden

ISSN 1104-1560
ISBN 91-554-3305-7

Cover design: Jerk-Olof Werkmäster
Printed in Sweden by Norstedts Tryckeri, Stockholm
Distributor: Almqvist & Wiksell International, Box 4627, S-116 91 Stockholm

CONTENTS

List of Tables

List of Figures

Acknowledgments

The research reported in this thesis has been supported by NUTEK (the Swedish National Board for Industrial and Technical Development) and the research committee of the University of Örebro.

Many thanks to docent Bengt Swensson for his way of pushing me to start work on this thesis, and for his hinting at the solutions to certain problems in such a way that I sometimes believed that I found them myself. I am also grateful to Bengt for his involvement in reading and discussing earlier versions of manuscripts, which has enabled me to greatly improve the clarity and contents of the thesis. Thank you, Bengt!

I also want to thank my supervisor professor Anders Christoffersson for his views and discussions of the manuscript of this thesis.

Professor Carl Erik Särndal is gratefully acknowledged for his views and comments on the final version of the manuscript. Valuable suggestions on the English text was made by Mr Malcolm Forbes.

I am also indebted to my colleagues at the University of Örebro for their friendly attitude to my work, and I am grateful for the admiration from my younger colleagues.

Last but not least I am grateful for the never-ending encouragement and patience on the part of my wife Kerstin and our children Jacob, Anna and Erik. Thank you for all of your time that you have given to me.

Örebro, April 1994
Claes Andersson

Summary

In many surveys, the units in the population constitute a number of domains, and comparisons are to be made between the domains regarding some response variables. When the variables that are used to classify the units into domains contain errors, misclassification of units will occur. Errors in the response variables that are used to compute totals, ratios or means within domains are called measurement errors. Although misclassification is a measurement error a distinction is made between them in this thesis.

Misclassification might for example occur when a fast, inexpensive but fallible device has been used to determine the correct domain of an individual instead of a lengthy, expensive but true device. The erroneous assignment of sample units can cause a systematic error in the estimation of a parameter that is a function of one or more response variables even when there are no measurement errors present.

In this thesis it is assumed that two measuring devices are available for measurement of the response variables and classification of units. The first, fallible device is a cheap procedure that tends to misclassify data but may or may not measure the response variables correctly. The second, true device is an expensive procedure that classifies units and measures the response variables correctly.

To estimate parameters of interest in different domains an initial sample of n_a units is taken and the fallible device is used. Possibly $n_a=N$, i.e. the initial sample is a census. A subsample of n units is drawn from the initial sample and the true device is used in the subsample to get true classifications and measurements. For each unit in the subsample we have data about the classifications and the measurements from the true and fallible devices.

When the fallible data in the usually large initial sample are used, we will get estimates that are affected by non-sampling errors but have small sampling errors. On the other hand when the data from the usually small subsample are used we will get estimates without non-sampling errors with large sampling errors.

The information in the error-prone data from the initial sample can be used to improve the estimates from the subsample. The generalised regression estimator, GREG, can be used for this purpose but it must be modified to cover the case with misclassification and measurement errors in the data from the first-phase sample.

Such a modified estimator, called *m*GREG, is presented in this thesis. It is shown that some of the estimators that have been proposed in the two-phase sampling situation described above can be considered as special cases of the *m*GREG estimator.

The *m*GREG estimator can, as opposed to other proposed estimators, be used with arbitrary sample designs in the first and second phases, i.e. not only when simple random sampling (srs) is used. Moreover, it can be used to estimate any rational function of totals, i.e. not only means.

The effect of misclassification on the sample estimate of the binomial proportions was studied by Bross (1954) under the assumption of a probabilistic misclassification model. Bross' work was extended to multinomial data by Tenenbein (1970, 1971, 1972).

The estimation of domain means of a response variable when the sample units are allocated to domains by a fallible device was studied by Mak (1988) and Mak & Li (1988). The response variable is assumed to be measured without error when the fallible devices are used, while Selén (1986) allows errors in the response variable.

The problem of the allocation of the second-phase sample to get efficient estimates in all domains was treated by Deming (1977) and Haitovsky & Rapp (1992).

In *chapter 1* of this thesis, basic concepts like survey, survey errors, measurement errors, misclassification and a model for the simultaneous occurrence of measurement errors and misclassification are introduced.

The theory for two-phase sampling and the GREG estimator is described in *chapter 2*. In *chapter 3* the modification of GREG is presented. The modification is necessary since GREG assumes fixed values of the auxiliary variables. However, when misclassification and measurement errors are present, they are stochastic.

The estimation of domain means and relative domain sizes when data from the fallible device contain classification errors only, is considered in *chapter 4*, where also earlier work on the subject is reviewed. A Monte Carlo study is carried out to study the properties of the proposed estimators when srs is used in both phases.

A second Monte Carlo study is carried out for the case when the data from the fallible device contain misclassification and measurement errors. The study is presented in *chapter 5*, where a modified Mak & Li estimator is compared with *m*GREG when srs is used in both phases.

In *chapter 6* the *m*GREG estimator is developed for some commonly used designs, such as stratified sampling, cluster sampling and two-stage sampling.

In *chapter 7* an application of the *m*GREG estimator in the 1990 Swedish Population and Housing Census is presented.

The allocation problem is considered in *chapter 8*, where some proposed methods for the solution of the multicharacter problem in two-phase sampling are compared when srs is used in the first phase and stratification with srs within strata is used in the second phase.

The main results of the Monte Carlo studies in this thesis are that the *m*GREG estimator behaves well regarding bias, variance estimates and coverage rate of the estimated confidence intervals. Other proposed estimators must be rejected due to large bias or improper estimates of variances, which leads to coverage rates of the estimated confidence intervals that are too small.

CHAPTER 1

The Problem

1.1 Introduction

In many surveys, the finite population units constitute a set of disjoint subgroups or classes, and comparisons are to be made between these subgroups with respect to certain response variables. There may be errors in the variables that are used to classify the units into classes or subgroups, and these errors will cause misclassification. There may also be errors in the response variables that are used to compute totals, ratios, means or other parameters within classes or subgroups; these errors are called measurement errors. Although the outcomes of misclassification can be considered as variable values that are affected by measurement errors, a distinction between misclassification and measurement errors, in the above sense, will be made in this thesis.

Misclassification may for example occur when the determination of the correct class of an individual would require a lengthy and expensive process, and a fast, inexpensive but fallible device is used. The erroneous assignment of sample units to one category when they actually belong to another will, in general, make the estimator of an unknown parameter biased even if the response variables are measured without errors.

In this thesis it is assumed that two measuring devices are available to classify units and to measure the response variables. The first, true, device is expensive and/or difficult to use but it classifies units correctly and, as well, measures the response variables correctly. The second, fallible, device is "quick and dirty" and tends to misclassify data. Furthermore, it may or may not measure the response variables correctly.

One example of data produced by a fallible device are data in administrative registers. Such data are often used as complements to the data that are collected directly from the units in a sample survey. Now, administrative data are usually collected for a specific purpose with no consideration of their possible use for other purposes. Definitions and coverage do not for example, in general exactly meet the needs of the secondary user, e.g. a researcher or a statistical agency, and hence the accuracy of the data may not be sufficient for this user's aims.

When administrative data are used for secondary purposes, one should adjust the estimates for errors of classification as well as for measurements and other non-sampling errors if it is assumed that the effects of these errors on the estimates are unacceptably large. To be able to make the necessary adjustments one usually has to collect additional information. A natural way to collect this information is, as pointed out by Ekholm & Palmgren (1987), to use a two-phase sampling scheme.

The situation may be described in the following way. 1) A random sample of size n_a is drawn from the population of interest according to a specific sampling design. Using the fallible device F, each unit in the sample is (*i*) classified into one of D classes as well as (*ii*) measured, possibly with error, with respect to a response variable of interest. Let n_{ad} be the number of units classified into class d (d=1, ..., D), and let $\hat{t}_{ad}$ be the estimate of t_d, the class d population total of the response variable values, where $\hat{t}_{ad}$ is based on the n_{ad} possibly error-prone measurements of the response variable. The result is:

Fallible device F Class 1	·	d	·	D	Sum
n_{a1}, $\hat{t}_{a1}$	·	n_{ad}, $\hat{t}_{ad}$	·	n_{aD}, $\hat{t}_{aD}$	n_a, $\hat{t}_a$

2) A random sub-sample of size n is drawn from the primary sample in 1) above according to a specific second-phase sampling design. Each unit in the second-phase sample is classified into one of D classes as well as measured, without error, with respect to the response variable by the use of the *true* device T. Let $n_{d\cdot}$ denote the number of units belonging to domain d. Now each of the n units in the second-phase sample is classified according to device F as well as device T, and we can use the true values of the variable of interest to estimate t_d. We have the following result:

	Domain	Fallible device F Class 1	·	d	·	D	Sum
	1	n_{11}, $\hat{t}_{11}$	·	n_{1d}, $\hat{t}_{1d}$	·	n_{1D}, $\hat{t}_{1D}$	$n_{1\cdot}$, $\hat{t}_{1\cdot}$
True	·	·	·	·	·	·	·
Device T	d	n_{d1}, $\hat{t}_{d1}$	·	n_{dd}, $\hat{t}_{dd}$	·	n_{dD}, $\hat{t}_{dD}$	$n_{d\cdot}$, $\hat{t}_{d\cdot}$
	·	·	·	·	·	·	·
	D	n_{D1}, $\hat{t}_{D1}$	·	n_{Dd}, $\hat{t}_{Dd}$	·	n_{DD}, $\hat{t}_{DD}$	$n_{D\cdot}$, $\hat{t}_{\cdot D}$
	Sum	$n_{\cdot 1}$, $\hat{t}_{\cdot 1}$	·	$n_{\cdot d}$, $\hat{t}_{\cdot d}$	·	$n_{\cdot D}$, $\hat{t}_{\cdot D}$	$n_{\cdot\cdot}$, $\hat{t}_{\cdot\cdot}$

The estimator $\hat{t}_{d.}$ of t_d does not suffer from bias due to measurement or classification errors, while $\hat{t}_{.d}$ suffers from bias due to misclassification only, and $\hat{t}_{ad}$ suffers from bias due to misclassification and, possibly, measurement errors.

By using a two-phase sampling scheme one collects the true values from a subsample of the initial sample by applying the true device. This means that for units in the subsample, values from the true as well as the fallible device are available, while for units in the non-subsampled part only values from the fallible device are available.

Two-phase sampling is a method that has been suggested for several different purposes in the literature, e.g. to adjust parameter estimates for bias due to nonresponse, and it is a general method that can be used for the adjustment of bias due to other non-sampling errors as well, for example misclassification and measurement errors.

Concerning when the parameter of interest is the number of units in different domains, the estimation problem has been treated (*i*) by Bross (1954) in the case with two domains and (*ii*) by Tenenbein (1970, 1971, 1972), who presented methods for the estimation of the true number of units and misclassification probabilities in the case with two or more domains by using two-phase sampling. The binomial case corresponds to the situation considered by Bross (1954).

Two-phase sampling schemes were also applied for the testing of independence in contingency tables by Chiacchierini & Arnold (1977), while Hochberg (1977) and Chen (1979) treated models for categorical data more generally. Ekholm & Palmgren (1987) modelled the probabilities of misclassification and incorporated them into a model with binary dependent variables and a number of explanatory factors. Swensen (1988) used the scheme in combination with a true and a fallible classifier to estimate the change in a proportion between two points in time.

Methods to estimate the subgroup means of a response variable when the units in a simple random sample are allocated to subgroups or classes by a fallible device were developed by Selén (1986), Mak (1988) and Mak & Li (1988) by using a double sampling scheme. The use of these estimators presumed that the response variable was measured without error when both the correct and the fallible devices were used. However, Selén (1986) has suggested an estimator that allows misclassification and errors in the response variable when the fallible device is used. The estimators suggested by Selén and Mak need the assumption that the probabilities of misclassification are unaffected by the values of the response variable.

In all the cases referenced above the sampling design in phase one is assumed to be simple random sampling with replacement, here denoted *SIR*, and in the second phase simple random sampling without replacement, here denoted *SI*.

In this thesis the performance of the suggested estimators is compared with that of a new, general estimator, here called *m*GREG. This estimator is a modification of the GREG estimator introduced by Särndal & Swensson (1987) and further elaborated by Särndal, Swensson & Wretman (1992). The modified estimator can be used for the estimation of the number of units in a domain, the total of a response variable, the mean or the ratio in a subgroup, or some other function of totals, when data contain misclassification and/or measurement errors and two-phase sampling is used to collect information on the pattern of misclassification.

Unlike the estimators proposed by Mak and Mak & Li, the *m*GREG estimator allows measurement errors in the response variable when the fallible device is used, and unlike the estimators proposed by Selén and Mak no assumption of zero-correlation between classification errors and the response variable is needed. The *m*GREG estimator is also general in the sense that it can handle arbitrary sample designs, i.e. not only the *SIR* and *SI* designs.

In practice it is seldom possible to get correct information about every unit in the initial sample, which may even be a census. Then the user has two primary possibilities: 1) accept the probably biased estimates from the initial sample, and 2) adopt a strategy to adjust for the bias.

If the user chooses the second alternative, he may use two-phase sampling and apply the true device to the units in the second-phase sample to get correct values of the variables of interest. Then we have to distinguish between two situations.

In the first situation the initial sample is taken without consideration of the second phase because the user has no intention to take a subsample since he believes that his procedures are correct. However, during the data collection process he becomes aware that this is not the case, and he decides to take a subsample to get correct values. This is probably the most common situation in practice.

In the second situation the user knows from the beginning that his procedures are error-prone, at least for some variables of interest, and he decides to use a second-phase sample to handle the situation. In this situation at least two questions arise: 1) for what proportion of units should correct values be collected when resources are limited and precise information is expensive, and 2) how should the samples be allocated in the most efficient way?

When it comes to estimation, the user can choose to use only the correct data from the subsample or to select an estimator that uses both the correct data and the information (if any) about the true values in the error-prone data. By a combination of the information in the true and the error-prone data it may be possible to get better estimates of the parameters of interest. These matters are discussed in the final section of the thesis.

1.2 Population Characteristics

Let $U=\{1, \ldots, k, \ldots, N\}$ be a finite population of N units. This population is partitioned into D non-overlapping subsets, $U_1, \ldots, U_d, \ldots, U_D$, sometimes called subgroups or domains. Let N_d be the number of units in domain d. We have the partitioning equations

$$U = \bigcup_{d=1}^{D} U_d \quad ; \quad N = \sum_{d=1}^{D} N_d \tag{1.1}$$

For the dth domain, define

$$\zeta_{dk} = \begin{cases} 1 \text{ if } k \in U_d \\ 0 \text{ otherwise} \end{cases} \tag{1.2}$$

The vector

$$\zeta_k=(\zeta_{1k}, \ldots, \zeta_{dk}, \ldots, \zeta_{Dk})' \tag{1.3}$$

consists of D-1 "0" and a single entry "1" specifying the domain of the kth unit. The domain size can be expressed as

$$N_d = \Sigma_U \zeta_{dk} \tag{1.4}$$

NOTE: In (1.4) above, $\Sigma_U \zeta_{dk}$ is abbreviated notation for $\Sigma_{k \in U} \zeta_{dk}$. More generally, if A is any set of population units such that $A \subseteq U$, Σ_A is the notation for $\Sigma_{k \in A}$. □

Let y denote the response variable, and let y_k be the value of y for the kth population unit. The total of y in the population is denoted

$$t = \Sigma_U y_k \tag{1.5}$$

and the population mean of y is denoted

$$\bar{y}_U = \frac{t}{N} \tag{1.6}$$

I am especially interested in the total and the mean of y in domain d,

$$t_d = \Sigma_U \zeta_{dk} y_k = \Sigma_{U_d} y_k \tag{1.7}$$

and

$$\bar{y}_d = \frac{t_d}{N_d} \tag{1.8}$$

If the variable y_{dk} is defined as

$$y_{dk} = y_k \zeta_{dk} = \begin{cases} y_k \text{ if } k \in U_d \\ 0 \text{ otherwise} \end{cases} \tag{1.9}$$

then (1.7) can be written $t_d = \Sigma_U y_{dk}$.

1.3 Estimating t_d and $\bar{y}_d$ under Perfect Conditions

Neither the vector ζ_k nor the value y_k is assumed to be known for unit $k \in U$. When there is no misclassification and no measurement error present, i.e. when we are able to use the true device, we observe ζ_k and y_k in a sample s_a from U to estimate the parameters of interest. The index a is used to identify the initial sample and is needed when the two-phase sampling approach is considered. The sample s_a of size n_a is drawn by a design denoted $p_a(\cdot)$, such that $p_a(s_a)$ is the probability of choosing s_a. The inclusion probabilities are defined by

$$\pi_{ak} = \Pr(k \in s_a) \;\; ; \;\; \pi_{akl} = \Pr(k \& l \in s_a) \tag{1.10}$$

with $\pi_{akk} = \pi_{ak}$. It is assumed that $\pi_{ak} > 0$ for $k \in U$ and $\pi_{akl} > 0$ for $k \neq l$, $k \& l \in U$.

Depending on the nature of available auxiliary information, many different estimators can be used to estimate t_d. One simple, and frequently used, estimator is the π estimator (see Särndal *et al.* 1992).

The total t_d is unbiasedly estimated by the π estimator by

$$\hat{t}_{ad} = \Sigma_{s_a} \frac{\zeta_{dk} y_k}{\pi_{ak}} = \Sigma_{s_a} \frac{y_{dk}}{\pi_{ak}} \tag{1.11}$$

The variance $V(\hat{t}_{ad})$ of $\hat{t}_{ad}$ is given by

$$V(\hat{t}_{ad}) = \Sigma \Sigma_U \Delta_{akl} \frac{y_{dk}}{\pi_{ak}} \frac{y_{dl}}{\pi_{al}} \tag{1.12}$$

where $$\Delta_{akl} = \pi_{akl} - \pi_{ak} \pi_{al} \tag{1.13}$$

An unbiased estimator of (1.12) is given by

$$\hat{V}(\hat{t}_{ad}) = \Sigma \Sigma_{s_a} \frac{\Delta_{akl}}{\pi_{akl}} \frac{y_{dk}}{\pi_{ak}} \frac{y_{dl}}{\pi_{al}} \tag{1.14}$$

The mean in domain d is estimated by

$$\hat{\bar{y}}_{ad} = \frac{\hat{t}_{ad}}{\hat{N}_{ad}} \tag{1.15}$$

where

$$\hat{N}_{ad} = \sum_{s_a} \frac{\zeta_{dk}}{\pi_{ak}} \tag{1.16}$$

The estimator (1.15) is not an unbiased estimator, but the bias is small when n_a is large.

Remark 1.3.1. The approximate variance of a function of two or more estimated totals, e.g. the mean in a domain, is here obtained by using the first-order Taylor series approximation of the function of estimated totals around their expected values. The variance is estimated by replacing the population totals with their sample estimates in the derivatives. The approximate variance of $\hat{\bar{y}}_{ad}$ above is then estimated by

$$\hat{V}(\hat{\bar{y}}_{ad}) = \frac{1}{\hat{N}_{ad}^2}\left\{\hat{V}(\hat{t}_{ad}) + \hat{\bar{y}}_{ad}^2 \hat{V}(\hat{N}_{ad}) - 2\hat{\bar{y}}_{ad}\hat{C}(\hat{t}_{ad}, \hat{N}_{ad})\right\} \tag{1.17}$$

where $\hat{C}(\hat{t}_{ad}, \hat{N}_{ad})$ is the estimated covariance of $\hat{t}_{ad}$ and $\hat{N}_{ad}$. □

1.4 Estimating t_d and $\bar{y}_d$ under Non-Perfect Conditions

Even when more precise classifications of units and/or measurements of the y-variable are possible there may be time and cost factors that make it necessary to use cruder classification and/or measurement methods in a large survey.

The observed values for a unit in the sample s_a are assumed to be the result of the *procedure* F. A *procedure* is defined as all the different elements of the data collection step, including data editing, imputation, etc. The procedure F will not always give the true values of y_k and the vector ζ_k for the kth unit. Instead of ζ_k we observe the value $\mathbf{z}_k=(z_{1k}, \ldots, z_{dk}, \ldots, z_{Dk})'$ of the stochastic vector $\mathbf{Z}_k=(Z_{1k}, \ldots, Z_{dk}, \ldots, Z_{Dk})'$ where $k\in s_a$. The elements of $\mathbf{z}_k$ are defined as

$$z_{dk} = \begin{cases} 1 \text{ if } k \rightarrow d \\ 0 \text{ otherwise} \end{cases} \tag{1.18}$$

where $k \rightarrow d$ means "unit k is allocated to *class* d", d=1, 2, ..., D. The vector $\mathbf{z}_k$ consists of D-1 "0" and one "1". If the procedure F gives the correct classification of all units, then $\mathbf{z}_k=\zeta_k$ for all $k\in s_a$ and all $s_a\in U$.

Remark 1.4.1. Usually classes 1 to D correspond to domains 1 to D. In practice this does not have to be the case. One obvious situation is when classes 1 to D correspond to domains 1 to D while class D+1 contains all units that were impossible to classify due to lack of information. Hochberg (1977) refers to another example where the classification device used in procedure F uses a different scale than the true device.□

When the observed value of y contains measurement errors we observe the value x_k of the stochastic variable X_k instead of the value y_k for the kth unit in the sample s_a. If there are no measurement errors in the y-variable then $x_k=y_k$ for all $k\in s_a$ and all $s_a \in U$.

The variable X_{dk} is defined as $X_{dk}=X_k Z_{dk}$ and the value x_{dk} as

$$x_{dk} = x_k z_{dk} = \begin{cases} x_k \text{ if } k \rightarrow d \\ 0 \text{ otherwise} \end{cases} \tag{1.19}$$

Let $\mathbf{x}_k$ be a vector of size D with element d equal to the observed value x_{dk}, i.e. the vector $\mathbf{x}_k$ contains D-1 "0" and one element x_k.

When the π estimator is used to estimate t_d and x_{dk} is used in place of y_{dk}, the properties of the estimator $\hat{t}_{axd} = \sum_{s_a} x_{dk}/\pi_{ak}$ depend on the combined effects of the misclassification and measurement errors in x_{dk}.

To discuss the statistical properties of $\hat{t}_{axd}$ we have to use a model of the measurement and classification errors.

1.5 A Model for Measurement Errors and Misclassification

A fixed value, $y_{dk}=\zeta_{dk}y_k$, is assumed to exist for every unit $k\in U$ as well as a fallible procedure F that produces error-prone values of the vector ζ and variable y. The errors in ζ_k and/or y_k are assumed to be random, and thus the observed values $x_{dk}=z_{dk}x_k$ will in general differ from the value y_{dk} in a random way.

Following Särndal *et al.* (1992), it is assumed that for a given sample s_a, the random vectors $(\mathbf{Z}_k, X_k)'$, $k\in s_a$, have a certain joint probability distribution conditional on s_a, called a measurement model and denoted m.

Suppose measurements of (ζ_k, y_k) could be independently repeated many times on a sample s_a, thus generating a long series of observations on each element $k\in s_a$. The observed values for a particular unit k may not necessarily be the same in all repetitions, but may vary in a random fashion around its "long run" mean value μ_{xdk}, and with the "long run" variance σ^2_{xdk}. Correspondingly, the measurements of two elements k and l ($\in s_a$) will have the "long run" covariances σ_{xdkl}. The first- and second-order moments are assumed not to depend on the sample s_a. Thus the model moments are defined for all units in U provided $\pi_{ak} > 0$ for $k \in U$ and $\pi_{akl} > 0$ for $k \neq l \in U$.

Consider the estimator of interest, $\hat{t}_{axd} = \sum_{s_a} x_{dk} / \pi_{ak}$. The expectation of $\hat{t}_{axd}$ under the design $p_a(\cdot)$ and the model m is

$$E_{p_a m}(\hat{t}_{axd}) = E_{p_a}\left(E_m(\hat{t}_{axd} | s_a)\right) = \sum_U \mu_{xdk} \tag{1.20}$$

The bias of $\hat{t}_{axd}$ is $\sum_U \mu_{xdk} - t_d$, and it may be noted that even if the "unit" bias $\mu_{xdk} - y_{dk}$ differs from zero and even if this bias might be substantial for a given unit $k \in s_a$, the "overall" bias of $\hat{t}_{axd}$ may be small due to compensation between units.

For the design-model variance of $\hat{t}_{axd}$ (see Wolter 1985) we have

$$\begin{aligned} V_{p_a m}(\hat{t}_{axd}) &= V_{p_a} E_m(\hat{t}_{axd} | s_a) + E_{p_a} V_m(\hat{t}_{axd} | s_a) = \\ &= \sum \sum_U \Delta_{akl} \frac{\mu_{xdk}}{\pi_{ak}} \frac{\mu_{xdl}}{\pi_{al}} + \sum_U \frac{\sigma^2_{xdk}}{\pi_{ak}} + \sum_{k \neq l} \sum_U \frac{\pi_{akl} \sigma_{xdkl}}{\pi_{ak} \pi_{al}} = \\ &= V_1 + V_{21} + V_{22} \end{aligned} \tag{1.21}$$

The first component V_1 is called the sampling variance. The second component V_{21} is usually referred to as "the simple measurement" variance. The third component V_{22} is usually referred to as "the correlated measurement" variance, which is zero if the measurements of units k and l are independent. The first component V_1 equals zero under complete enumeration of U, e.g. a census, while V_{21}+V_{22} will probably differ from zero when the observed values contain measurement errors.

For a certain procedure F and a given unit $k \in s_a$, the first- and second-order moments of Z_{dk} under m are

$$\mu_{zk} = \theta_{dk} = E_m(Z_{dk} | s_a) \tag{1.22}$$

$$\sigma^2_{zk} = \theta_{dk}(1 - \theta_{dk}) = V_m(Z_{dk} | s_a) \tag{1.23}$$

$$\sigma_{zkl} = \theta_{dkl} - \theta_{dk}\theta_{dl} = C_m(Z_{dk}, Z_{dl} | s_a) \tag{1.24}$$

where $\theta_{dkl} = E_m(Z_{dk} Z_{dl} | s_a)$.

The first- and second-order moments of X_k under m are

$$\mu_{xk} = E_m(X_k | s_a) \tag{1.25}$$

$$\sigma^2_{xk} = V_m(X_k | s_a) \tag{1.26}$$

$$\sigma_{xkl} = C_m(X_k, X_l | s_a) \tag{1.27}$$

The expected value μ_{zk} of the indicator variable Z_{dk} will differ from its true value 1 or 0 unless the measurement of ζ_k is always error-free, while the expected value μ_{xk} of X_k may or may not be equal to y_k. It follows that the expected value μ_{xdk} of X_{dk} will in general not be equal to its true value even if $\mu_{xk}=y_k$ when there are classification errors present.

Expressing the first-order moments of X_{dk} in terms of the first-order moments of X_k and Z_{dk}, we get

$$\mu_{xdk} = \mu_{xk}^{(1)}\theta_{dk} \tag{1.28}$$

where $\mu_{xk}^{(1)}$ is the conditional expectation of X_k given that $Z_{dk}=1$. Assuming that σ_{xk}^2 and σ_{xkl} do not depend on Z_{dk}, the variance of X_{dk} is

$$\sigma_{xdk}^2 = \sigma_{xk}^2\theta_{dk} + \left(\mu_{xk}^{(1)}\right)^2\sigma_{zk}^2 \tag{1.29}$$

and the covariance between X_{dk} and X_{dl} is

$$\sigma_{xdkl} = \theta_{dkl}\sigma_{xkl} + \mu_{xk}^{(1)}\mu_{xl}^{(1)}\sigma_{zkl} \tag{1.30}$$

Remark 1.5.1. The covariance between Z_{dk} and X_k is $\sigma_{zk}^2(\mu_{xk}^{(1)} - \mu_{xk}^{(0)})$, where $\mu_{xk}^{(0)}$ is the conditional expectation of X_k given that $Z_{dk}=0$. If X_k and Z_{dk} are uncorrelated, then $\mu_{xk}^{(1)} = \mu_{xk}^{(0)} = \mu_{xk}$. □

Both the expectation and the variance of $\hat{t}_{axd}$ are affected by the possible misclassification and/or measurement errors in each unit. The effects depend on the probability of misclassification, the proportion of elements actually belonging to the different domains and the correlation between the expected x_k-values and the θ_{dk}-values.

Koch, Freeman & Freeman (1975) have shown that the variance (1.21) will be underestimated when the measurements of units k and l are uncorrelated and the observed x_{dk} values are used in expression (1.14) as an estimator of $V(\hat{t}_{axd})$, the bias of the "standard" estimator (1.14) is then $-\Sigma_U \sigma_{xdk}^2$.

1.6 Implications of the Measurement Model

Four cases can be identified depending on the occurrence of misclassification and measurement errors when procedure F is used.
Note that "=" means equality for *every* k and "≠" means inequality for *some* k.

	$z=\zeta$	$z\neq\zeta$
$x=y$	I	II
$x\neq y$	III	IV

Case I does not pose any problems, since it can be handled by standard estimation methods.

Case II means that there are no measurement errors, only classification errors. This case was treated by Selén (1986), Mak (1988), Mak & Li (1988) and Andersson (1992). Selén and Mak assume that y_k and θ_{dk} are uncorrelated, while Andersson allows correlation between y_k and θ_{dk}. When y_k and θ_{dk} are uncorrelated for $d=1, 2, \ldots,$ D and $k\in U_{d'}$ ($d'=1, 2, \ldots, D$) we may write $\Sigma_{U_{d'}} y_k\theta_{dk} = \theta_{d|d'}t_{yd'}$, where

$$\theta_{d|d'} = \frac{1}{N_{d'}}\Sigma_{U_{d'}}\theta_{dk} \tag{1.31}$$

may be interpreted as the average risk of classifying an element that belongs to domain d' into class d. The expected value of the estimator $\hat{t}_{axd}$ under the design $p_a(\cdot)$ and model m can then be written

$$E_{p_a m}(\hat{t}_{axd}) = t_{yd} + \sum_{d'\neq d}^{D}(\theta_{d|d'}t_{yd'} - \theta_{d'|d}t_{yd}) = t_{yd} + Bias\,(\hat{t}_{axd}) \tag{1.32}$$

Selén (1986) showed that the expected value of the mean value $\hat{\bar{x}}_d$ is

$$E(\hat{\bar{x}}_d) \approx \frac{E(\hat{t}_{axd})}{E(\hat{N}_{ad})} = \bar{y}_d + \sum_{d'\neq d}^{D}\phi_{d'|d}(\bar{y}_{d'} - \bar{y}_d) \tag{1.33}$$

where $\phi_{d'|d}$ is the *a posteriori* probability $\Pr(k\in U_{d'} \mid k\rightarrow d)$.

Case II also covers the situation when the units are only classified into a number of classes. Then $\hat{t}_{axd} = \hat{N}_{ad}$ is the estimated number of units falling into class d and $\hat{p}_{ad} = \hat{N}_{ad} / N$ is the estimated proportion of units falling into class d. When the sample design is *SI*, the expected value of $\hat{p}_{ad}$ can be written

$$E(\hat{p}_{ad}) = \sum_{d'=1}^{D}\theta_{d|d'}P_{d'} \tag{1.34}$$

where $P_{d'} = N_{d'}/N$.

This situation has been treated when the sampling design is *SIR* by Bross (1954), Hansen, Hurwitz & Bershad (1961), Cochran (1968), Tenenbein (1970, 1971, 1972), Hochberg (1977) and Chen (1979).

Case III means that there are no classification errors, only measurement errors. A simple measurement model that can be used for a general design in this case is presented by Särndal, Swensson & Wretman (1992).

Case IV means that there are both classification errors and measurement errors present. This case has not been treated very extensively but Selén (1986) and Andersson (1992) have suggested different estimators.

Example. The estimator (1.11) of the total t_d, using the first-phase sample s_a when an *SI* sample is taken, takes the form

$$\hat{t}_{axd} = \frac{N}{n_a} \Sigma_{s_a} x_{dk} \, . \tag{1.35}$$

Using the model m and assuming case IV, define

$$S^2_{\mu_{xd}} = \frac{1}{N-1} \Sigma_U (\mu_{xdk} - \bar{\mu}_{xd})^2 \;\; ; \;\; \bar{\mu}_{xd} = \frac{1}{N} \Sigma_U \mu_{xdk} \tag{1.36}$$

The variance (1.21) of $\hat{t}_{axd}$ now can be written

$$V(\hat{t}_{axd}) = \frac{N^2}{n_a}\left(1 - \frac{n_a}{N}\right) S^2_{\mu_{xd}} + \frac{N}{n_a} \Sigma_U \sigma^2_{xdk} + \frac{n_a - 1}{n_a} \frac{N}{N-1} \sum_{k \neq l} \Sigma_U \sigma_{xdkl} = \tag{1.37}$$

$$= V_1 + V_{21} + V_{22}$$

where V_1 is the sampling variance and where V_{21} and V_{22} are the simple measurement variance and the correlated measurement variance, respectively, which can be written in the well-known forms

$$V_{21} = N^2 \frac{\sigma^2_{xd}}{n_a} \quad \text{and} \quad V_{22} = N^2 \frac{n_a - 1}{n_a} \rho_{xd} \sigma^2_{xd} \tag{1.38}$$

where σ^2_{xd} and $\rho_{xd}\sigma^2_{xd}$ are the averages of σ^2_{xdk} and σ_{xdkl} in U. When measurement errors are present, the variance of $\hat{t}_{axd}$ will increase if $V_{21}+V_{22}$ is positive, which is usually the case in practice. □

When the bias of the estimator of t_d or $\bar{y}_d$ is deemed too large to be acceptable when the procedure F is used, at least for some parameters, the user has the following choice: 1) find an acceptable procedure F_0 instead of F, or 2) use a true procedure T to get error-free data at least for a subsample of s_a. The second of these alternatives is usually referred to as two-phase or double sampling and has been used extensively in this situation.

Different approaches to estimate the number of units, or a mean in a domain, have been suggested to handle misclassification when the θ_{dk}s are unknown. Common to all these approaches is the use of repeated measurements of the selected units, either with the same procedure in each repetition or with a better but not necessarily perfect method. When the same procedure is used on each unit in every repetition, one repetition is sufficient when D=2. In the case D>2 at least two repetitions for each unit are necessary unless restrictions are imposed on θ_{dk}. Such methods have been suggested by Cochran (1968), Chua & Fuller (1987), Fuller (1986), Nathan & Eliav (1988) and Selén (1990). These methods, however, will not be treated here. Instead I will treat the estimation problem in the two-phase sampling situation outlined in the next chapter.

When the π estimator is used to estimate t_d with the correct values y_k and vectors ζ_k from the subsample, the estimates are free from bias but suffer from a large standard error when n is smaller than n_a. On the other hand, when the fallible values x_{dk} recorded in the sample s_a are used with the π estimator, the estimates are usually biased but will have a smaller standard error when n_a>n. If we combine the large but erroneous sample with the usually small sample with correct values, we ought to be able to improve the situation. One way of doing this will be described in the following chapters.

CHAPTER 2

Two-Phase Sampling and the GREG Estimator

2.1 Two-Phase Sampling

Two-phase sampling was introduced by Neyman (1938) and has been a standard technique in survey sampling for a long time. The technique has been used for several different purposes. For example Raj (1968) considers two-phase sampling for difference estimation, for pps-estimation, for ratio estimation, for simple regression estimation where x_k is observed in the first-phase sample and is a proxy for y_k that is observed in the second-phase sample, for stratification, as a tool for reducing nonresponse bias in survey results, etc.

The theory for two-phase sampling with arbitrary sampling designs in each phase and a regression-type estimator was developed by Särndal & Swensson (1987) and further elaborated by Särndal *et al.* (1992). They also developed a general approach for estimating the variance of a two-phase estimator under a general design.

Here, two-phase sampling is used to get correct, or at least significantly better, classification and/or measurements of a subsample of the units in the first-phase sample s_a.

A subsample s of size n from s_a is taken according to the design $p(\cdot \mid s_a)$. In the sample s the correct values of y_k and the vector ζ_k are observed by using a procedure T. Thus, for elements in the sample s, we have the correct values of the vector ζ_k and of y_k as well as the possibly fallible values $\mathbf{z}_k$ and x_k.

The first- and second-order inclusion probabilities of the units in the sample s, given the first sample s_a, are

$$\pi_{k|a}=\Pr(k\in s|s_a) \;\; ; \;\; \pi_{kl|a}=\Pr(k\&l\in s|s_a) \tag{2.1}$$

Define
$$\pi_k^* = \pi_{ak}\pi_{k|a} \;\; ; \;\; \pi_{kl}^* = \pi_{akl}\pi_{kl|a} \tag{2.2}$$

where $\pi_{kk}^* = \pi_k^*$. It is assumed that $\pi_k^* > 0$ and $\pi_{kl}^* > 0$ for $k\&l\in s_a$.

Further, let

$$\Delta_{kl|a} = \pi_{kl|a} - \pi_{k|a}\pi_{l|a} \tag{2.3}$$

and
$$\Delta_{kl}^* = \pi_{kl}^* - \pi_k^*\pi_l^* \tag{2.4}$$

Note that, in general, $\pi_k^* \neq \pi_k$=Pr($k \in s$). However, in the important situation when both $p_a(\cdot)$ and $p(\cdot \mid s_a)$ are *SI* designs, then $\pi_k^* = \pi_k$.

Since the correct values of y_k and the vector ζ_k are observed in the sample s, the total t_d can be estimated unbiasedly by using the π^* estimator (Särndal *et al.* 1992):

$$\hat{t}_d^* = \Sigma_s \frac{y_{dk}}{\pi_k^*} \tag{2.5}$$

The variance of $\hat{t}_d^*$ is given by

$$V(\hat{t}_d^*) = \Sigma\Sigma_U \Delta_{akl} \frac{y_{dk}}{\pi_{ak}} \frac{y_{dl}}{\pi_{al}} + E_{p_a}\left[\Sigma\Sigma_{s_a} \Delta_{kl|a} \frac{y_{dk}}{\pi_k^*} \frac{y_{dl}}{\pi_l^*}\right] \tag{2.6}$$

where Δ_{akl} is defined in (1.13).

The variance estimator can be written as a compact expression

$$\hat{V}(\hat{t}_d^*) = \Sigma\Sigma_s \frac{\Delta_{kl}^*}{\pi_{kl}^*} \frac{y_{dk}}{\pi_k^*} \frac{y_{dl}}{\pi_l^*} \tag{2.7}$$

In practice, it is realistic to assume that the sample size n of s is rather small compared to n_a, which means that the variance (2.6) will be fairly large.

When two-phase sampling is used to collect error-free measurements of the classification and measurement variables, it is also possible to estimate the variance component V_{21} in (1.21).

The simple response variance V_{21} can be estimated by the technique suggested by Biemer & Forsman (1992). They suggest the estimator

$$\hat{G}_d = \Sigma_s \frac{\left(x_{dk} - y_{dk}\right)^2}{\pi_{ak}\pi_k^*} \tag{2.8}$$

which, however, is a biased estimate of V_{21} since

$$\begin{aligned} E(\hat{G}_d) &= E_{p_a} E_p E_m\left(\hat{G}_d \middle| s_a\right) = \\ &= \Sigma_U \frac{\sigma_{xdk}^2}{\pi_{ak}} + \Sigma_U \frac{\left(y_{dk} - \mu_{xdk}\right)^2}{\pi_{ak}} = V_{21} + B_{21} \end{aligned} \tag{2.9}$$

When $p_a(\cdot)$ and $p(\cdot \mid s_a)$ are *SI* designs the expectation of $\hat{G}_d$ is given by

$$E(\hat{G}_d) = \frac{N}{n_a} \Sigma_U \left\{\sigma_{xdk}^2 + \left(y_{dk} - \mu_{xdk}\right)^2\right\}. \tag{2.10}$$

The bias term B_{21} in (2.9) is neither known nor possible to estimate unbiasedly. Biemer & Forsman (1992) claim that if the responses of individuals over repeated trials are on average equal to the true values, the bias B_{21} will be small. However, when estimation in domains is of interest this may not be true in the presence of misclassification. In designs where π_{ak} differs between units, e.g. stratified designs and clustered designs, it is important to note that large errors in units with small inclusion probabilities will cause great damage both to the variance of $\hat{t}_{axd}$ and to the estimate of V_{21}. It is not unusual in practice that units which are considered less important are given small inclusion probabilities as well as little attention with respect to measurement problems.

The presence of doubly classified units makes it possible to use the data from the second-phase sample in combination with the error-prone data in the first-phase sample to construct an approximately unbiased estimator of t_d with a smaller standard error than an estimator which only uses data for the units in the second-phase sample. This is possible if the y_{dk}-values and the x_{dk}-values are correlated, as they should be. If they are not correlated it means that procedure F is useless in practice.

2.2 The GREG Estimator — A General Description

When there are auxiliary variables available before the sampling in the second phase that are related to the true values y_{dk}, a better estimator than $\hat{t}_{axd}$ may be constructed that combines approximate unbiasedness with a small variance. A general regression estimator, GREG, that can be used in many useful situations and for arbitrary sampling designs when auxiliary information exists, has been developed by Särndal & Swensson (1987).

The GREG estimator presumes fixed auxiliary variable values. In this chapter the GREG estimator is described under this presumption. Three features are considered: 1) domain estimates, 2) two-phase sampling and 3) the GREG estimator. In Särndal *et al.* (1992) explicit results are given for the combinations {1, 3} and {2, 3}. In this chapter explicit results are given for the combination of all three features. In the next chapter the GREG estimator in combination with domain estimates in two-phase sampling is modified to fit the situation when the values of the auxiliary variables are stochastic as is the case in the situation described in chapter 1.

In the two-phase sampling context two situations are distinguished corresponding to the nature of the auxiliary information.

i) The vector $\mathbf{x}_1$ contains J_1 auxiliary variables and the *fixed* value $\mathbf{x}_{1k}$ is known for all units in the population U.

ii) The vector $\mathbf{x}=(\mathbf{x}'_1, \mathbf{x}'_2)'$ contains $J=J_1+J_2$ auxiliary variables and the *fixed* value $\mathbf{x}_k=(\mathbf{x}'_{1k}, \mathbf{x}'_{2k})'$ is known for all units in the sample s_a.

For a full account of the GREG estimator in the double sampling case, see Särndal *et al.* (1992). Here, only the case where $\mathbf{x}=\mathbf{x}_2$ and $J=J_2=D$ will be considered, i.e. where auxiliary information is available for the units in the first sample s_a only.

Again, note that the auxiliary variables are assumed to have *fixed* values for each unit k, contrary to the situation described in chapter 1 above.

The GREG estimator is motivated as follows. Assume that the scatter of the N points $\{(y_{dk}, x_{1k}, \ldots, x_{Dk}) : k=1, \ldots, N\}$ looks *as if* it had been generated by a linear regression model called ξ with y_d as the regressor and $\mathbf{x}=(x_1, \ldots, x_D)'$ as the regressands.

The regression model ξ has the following features:

i) $y_{d1}, \ldots, y_{dk}, \ldots, y_{dN}$ are assumed to be realised values of independent random variables $Y_{d1}, \ldots, Y_{dk}, \ldots, Y_{dN}$,

ii) $E_\xi(Y_{dk}) = \mathbf{x}'_k \boldsymbol{\beta}_d = x_{1k}\beta_{d1}+\cdots+x_{Dk}\beta_{dD}$ $\qquad (k=1, \ldots, N)$,

iii) $V_\xi(Y_{dk}) = \sigma^2_{dk}$ $\qquad (k=1, \ldots, N)$,

where E_ξ and V_ξ denote expected value and variance with respect to the model ξ while $\boldsymbol{\beta}_d$ and σ^2_{dk} are model parameters. Here, the σ^2_{dk} are known up to multiplicative constants that vanish when $\boldsymbol{\beta}_d$ is estimated.

The randomness introduced by ξ exists *only* by assumption and has nothing to do with the randomness created by sample selection through the sampling designs $p_a(\cdot)$ and $p(\cdot|s_a)$.

The role of the model ξ is to describe the point scatter in the finite population U. It is hoped that the model ξ fits the population reasonably well. However, the assumption is never made that the population was really generated by ξ. The model serves merely as a *tool* to derive an appropriate regression estimator, and the efficiency of the regression estimators compared to the π^* estimator will depend on the goodness of the fit. The basic properties of the regression estimator—approximate unbiasedness, validity of variance formulas etc.—are not dependent on whether the model holds or not. The estimator is thus model-*assisted* but it is not model-*dependent.*

If the y_{dk}-values were known for the whole sample s_a, an estimator of the unknown $\boldsymbol{\beta}_d$ vector could be formed

$$\mathbf{b}_{ad} = \left(\Sigma_{s_a} \frac{\mathbf{x}_k \mathbf{x}'_k}{\sigma^2_{dk}\pi_{ak}} \right)^{-1} \Sigma_{s_a} \frac{\mathbf{x}_k y_{dk}}{\sigma^2_{dk}\pi_{ak}} \tag{2.11}$$

and one could obtain the residuals

$$E_{dk} = y_{dk} - \mathbf{x}'_k \mathbf{b}_{ad} \qquad k \in s_a \tag{2.12}$$

What we can actually calculate from the available data when two-phase sampling is used is the regression coefficient vector $\mathbf{b}_d$

$$\mathbf{b}_d = \left(\sum_s \frac{\mathbf{x}_k \mathbf{x}'_k}{\sigma^2_{dk} \pi^*_k} \right)^{-1} \sum_s \frac{\mathbf{x}_k y_{dk}}{\sigma^2_{dk} \pi^*_k} \tag{2.13}$$

the predicted values

$$\hat{y}_{dk} = \mathbf{x}'_k \mathbf{b}_d \qquad k \in s_a \tag{2.14}$$

and the residuals

$$e_{dk} = y_{dk} - \mathbf{x}'_k \mathbf{b}_d \qquad k \in s \tag{2.15}$$

Let the vector of estimated totals from the first-phase sample be

$$\hat{\mathbf{t}}_{ax} = \sum_{s_a} \frac{\mathbf{x}_k}{\pi_{ak}} \tag{2.16}$$

and let the estimate of the same total from the second-phase sample be

$$\hat{\mathbf{t}}_x = \sum_s \frac{\mathbf{x}_k}{\pi^*_k} \tag{2.17}$$

The GREG estimator in the two-phase sampling case is formally defined as

$$\hat{t}_{dr} = \hat{t}^*_d + (\hat{\mathbf{t}}_{ax} - \hat{\mathbf{t}}_x)' \mathbf{b}_d \tag{2.18}$$

where $\mathbf{b}_d = (b_{d1}, \ldots, b_{dD})'$ is a vector of D estimated regression coefficients and $\hat{t}^*_d$ is defined by (2.5)

The estimator (2.18) above can also be expressed as

$$\hat{t}_{dr} = \sum_{s_a} \frac{\hat{y}_{dk}}{\pi_{ak}} + \sum_s \frac{e_{dk}}{\pi^*_k} \tag{2.19}$$

or as

$$\hat{t}_{dr} = \sum_s \frac{g_k y_{dk}}{\pi^*_k} \tag{2.20}$$

where

$$g_k = 1 + (\hat{\mathbf{t}}_{ax} - \hat{\mathbf{t}}_x)' \left(\sum_s \frac{\mathbf{x}_k \mathbf{x}'_k}{\sigma^2_{dk} \pi^*_k} \right)^{-1} \frac{\mathbf{x}_k}{\sigma^2_{dk}} \tag{2.21}$$

If there exists a constant column vector $\boldsymbol{\lambda}_d$ of dimension D that does not depend on k such that $\sigma^2_{dk} = \boldsymbol{\lambda}'_d \mathbf{x}_k$ for $k \in s_a$, then $\sum_s e_{dk} / \pi^*_k = 0$ for all samples s from s_a, and the GREG estimator can be written in the simple form, $\hat{t}_{dr} = \hat{\mathbf{t}}'_{ax} \mathbf{b}_d$.

The approximate variance of $\hat{t}_{dr}$ can be shown to be

$$AV(\hat{t}_{dr}) = \sum\sum_U \Delta_{akl} \frac{y_{dk}}{\pi_{ak}} \frac{y_{dl}}{\pi_{al}} + E_{p_a}\left[\sum\sum_{s_a} \Delta_{kl|a} \frac{E_{dk}}{\pi_k^*} \frac{E_{dl}}{\pi_l^*}\right] \tag{2.22}$$

An estimator of the variance of $\hat{t}_{dr}$ is suggested by Särndal *et al.* (1992) as

$$\hat{V}(\hat{t}_{dr}) = \sum\sum_s \frac{\Delta_{akl}}{\pi_{kl}^*} \frac{y_{dk}}{\pi_{ak}} \frac{y_{dl}}{\pi_{al}} + \sum\sum_s \frac{\Delta_{kl|a}}{\pi_{kl|a}} \frac{g_k e_{dk}}{\pi_k^*} \frac{g_l e_{dl}}{\pi_l^*} \tag{2.23}$$

with g_k=1 or with g_k defined by (2.21).

CHAPTER 3

The *m*GREG Estimator

3.1 A Modification of the GREG Estimator

The GREG estimator was developed under the assumption that the values of the auxiliary variable x are *fixed* in the population, i.e. that if a large number of samples are generated under the design $p_a(\cdot)$ the same value x_k is recorded every time a specific unit k is included in the sample.

In the situation with misclassification and/or measurement error described as in chapter 1 it is natural to use the recorded values x_k and z_{dk} in the sample s_a as auxiliary information in the estimation of the total t_d. If the values x_k and z_{dk} were fixed then the GREG estimator could be used. The estimator should be a good one since the results produced by procedures F and T ought to be highly correlated. However, the procedure F will probably produce stochastic values x_k for unit $k \in s_a$, i.e. the values are likely to vary in a random way if the procedure F is repeated for the same unit. In this case the assumptions for the GREG estimator are not met. The GREG estimator has to be slightly modified to cover the situation with stochastic auxiliary information.

It is well-known from the theory of regression analysis that the ordinary estimator for $\boldsymbol{\beta}_d$ is biased and even inconsistent when the regressands contain errors. However, our interest is *not* to estimate $\boldsymbol{\beta}_d$, our interest is to predict the y-values in the sample s_a. This can be done by estimating $\boldsymbol{\beta}_d$ by using the possibly erroneous x_k-values and then using the "conditional" form of the model ξ below to arrive at estimates of t_d and the approximate variance $AV(\hat{t}_d)$.

For a given sample s_a and a given realisation $\mathbf{x}_k$, $k \in s_a$ of the stochastic vector $\mathbf{X}_k$ whose first- and second-order moments are assumed to exist, the model ξ is formulated as

$$\begin{cases} E_\xi(y_{dk}|s_a, \mathbf{x}_k) = \mathbf{x}'_k \boldsymbol{\beta}_d \\ V_\xi(y_{dk}|s_a, \mathbf{x}_k) = \sigma^2_{dk} \end{cases} \tag{3.1}$$

With this formulation of the model ξ the modified GREG estimator, here called the *m*GREG estimator, of t_d looks formally the same as the GREG estimator

$$\hat{t}_{dr} = \hat{t}_d^* + (\hat{\mathbf{t}}_{ax} - \hat{\mathbf{t}}_x)' \, \mathbf{b}_d = \Sigma_s \frac{g_k y_{dk}}{\pi_k^*} \tag{3.2}$$

where $\mathbf{b}_d$, $\hat{\mathbf{t}}_{ax}$, $\hat{\mathbf{t}}_x$ and g_k are given by (2.13), (2.16), (2.17) and (2.21) in section 2.2.

It is easy to show that

$$\hat{t}_{dr} = \Sigma_{s_a} \frac{y_{dk}}{\pi_{ak}} + \Sigma_s \frac{g_k E_{dk}}{\pi_k^*} - \Sigma_{s_a} \frac{E_{dk}}{\pi_{ak}} \tag{3.3}$$

where E_{dk} is defined in (2.12), but

$$\Sigma_s \frac{g_k E_{dk}}{\pi_k^*} = \Sigma_s \frac{E_{dk}}{\pi_k^*} + (\hat{\mathbf{t}}_{ax} - \hat{\mathbf{t}}_x)' \, (\mathbf{b}_d - \mathbf{b}_{ad}) \tag{3.4}$$

Then

$$\hat{t}_{dr} - t_d = \Sigma_{s_a} \frac{y_{dk}}{\pi_{ak}} - \Sigma_U \, y_{dk} + \Sigma_s \frac{E_{dk}}{\pi_k^*} - \Sigma_{s_a} \frac{E_{dk}}{\pi_{ak}} + \Delta \tag{3.5}$$

where $\Delta = (\hat{\mathbf{t}}_{ax} - \hat{\mathbf{t}}_x)' \, (\mathbf{b}_d - \mathbf{b}_{ad})$ is small compared with the terms that precede.

The expected value of $\hat{t}_{dr} - t_d$ is approximately equal to zero since

$$E\left(\hat{t}_{dr} - t_d\right) = E_{p_a} E_{\mathbf{X}} E_p\left(\hat{t}_{dr} - t_d \middle| s_a, \mathbf{x}\right) \approx 0 \tag{3.6}$$

where $\mathbf{x}$ is an $n_a \times D$ matrix with row k equal to $\mathbf{x}'_k$. Thus $\hat{t}_{dr}$ is an approximately unbiased estimator of t_d.

The approximate variance of $\hat{t}_{dr}$ can be written as

$$\begin{aligned} AV\left(\hat{t}_{dr}\right) &= AV\left(\hat{t}_{dr} - t_d\right) = V_{p_a} E_{\mathbf{X}} E_p\left(\hat{t}_{dr} - t_d \middle| s_a, \mathbf{x}\right) + \\ &\quad + E_{p_a} V_{\mathbf{X}} E_p\left(\hat{t}_{dr} - t_d \middle| s_a, \mathbf{x}\right) + E_{p_a} E_{\mathbf{X}} V_p\left(\hat{t}_{dr} - t_d \middle| s_a, \mathbf{x}\right) \approx \\ &\approx \Sigma \Sigma_U \, \Delta_{akl} \frac{y_{dk}}{\pi_{ak}} \frac{y_{dl}}{\pi_{ak}} + 0 + E_{p_a} E_{\mathbf{X}} \left[\Sigma \Sigma_{s_a} \Delta_{kl|a} \frac{E_{dk}}{\pi_k^*} \frac{E_{dl}}{\pi_l^*} \right] \end{aligned} \tag{3.7}$$

The variance is estimated by

$$\hat{V}(\hat{t}_{dr}) = \Sigma \Sigma_s \frac{\Delta_{akl}}{\pi_{kl}^*} \frac{y_{dk}}{\pi_{ak}} \frac{y_{dl}}{\pi_{al}} + \Sigma \Sigma_s \frac{\Delta_{kl|a}}{\pi_{kl|a}} \frac{g_k e_{dk}}{\pi_k^*} \frac{g_l e_{dl}}{\pi_l^*} \tag{3.8}$$

with g_k=1 or with g_k as in (2.21) and e_{dk} defined as in (2.15), i.e. the variance estimators for the *m*GREG and GREG estimators of t_d are formally the same.

As with the GREG estimator, simplification of the estimator (3.2) occurs if the variance structure satisfies $\sigma^2_{dk}=\lambda'_d\mathbf{x}_k$ for all $k\in s_a$ and some constant vector λ_d not depending on k. One useful variance structure is

$$\sigma^2_{dk} = \sigma^2_d\mathbf{j}'\,\mathbf{x}_k \tag{3.9}$$

where σ^2_d is assumed to be an unknown constant and $\mathbf{j}'$ is a vector of length D with all elements equal to 1.

A number of estimators can now be derived from the general model. Remember that in the situation with misclassification and measurement errors the vector $\mathbf{x}_k$ contains only one non-zero element. Thus when $(k \in U_d)\&(k \to d')$ the variance structure (3.9) is $\sigma^2_{dk} = \sigma^2_d x_{d'k}$ and the matrix $\mathbf{x}_k\mathbf{x}'_k$ in (2.13) is diagonal with one diagonal element d' equal to $x^2_{d'k}$ and the other elements equal to zero. Then g_k is,

$$g_k = \begin{cases} \dfrac{\hat{t}_{axd'}}{\hat{t}^*_{xd'}} & \text{if } k \to d' \\ 0 & \text{otherwise} \end{cases} \tag{3.10}$$

and

$$\hat{t}_{dr} = \sum\nolimits_s \frac{g_k y_{dk}}{\pi^*_k} = \sum_{d'=1}^{D} \frac{\hat{t}_{axd'}}{\hat{t}^*_{xd'}}\hat{t}^*_{dd'} \tag{3.11}$$

where

$$\hat{t}^*_{dd'} = \sum\nolimits_s \frac{z_{d'k}\zeta_{dk}y_k}{\pi^*_k} \tag{3.12}$$

3.2 *SI* Designs in Both Phases — A Special Case

Let us discuss the simple situation when $p(\cdot)$ as well as $p_a(\cdot\,|\,s_a)$ are *SI* designs. Then the inclusion probabilities are given by

$$\pi_{ak} = \frac{n_a}{N},\quad k\in U \quad ;\quad \pi_{akl} = \frac{n_a}{N}\frac{n_a-1}{N-1},\quad k\neq l\in U \tag{3.13}$$

$$\pi_{k|a} = \frac{n}{n_a},\quad k\in s_a \quad ;\quad \pi_{kl|a} = \frac{n}{n_a}\frac{n-1}{n_a-1},\quad k\neq l\in s_a \tag{3.14}$$

$$\pi^*_k = \frac{n}{N},\quad k\in U \quad ;\quad \pi^*_{kl} = \frac{n}{N}\frac{n-1}{N-1},\quad k\neq l\in U \tag{3.15}$$

Assume that the variance structure (3.9) is $\sigma_{dk}^2 = \sigma_d^2 x_{dk}$ where σ_d^2 is an unknown constant. The element d' in $\mathbf{b}_d$ is then given by

$$b_{dd'} = \frac{\hat{t}_{dd'}^*}{\hat{t}_{xd'}^*} \tag{3.16}$$

where

$$\hat{t}_{dd'}^* = \frac{N}{n} \Sigma_s z_{d'k} \zeta_{dk} y_k \tag{3.17}$$

and

$$\hat{t}_{xd'}^* = \frac{N}{n} \Sigma_s z_{d'k} x_k \tag{3.18}$$

The *m*GREG estimator of t_d now takes the simple form

$$\hat{t}_{dr} = \sum_{d'=1}^{D} \hat{t}_{axd'} \frac{\hat{t}_{dd'}^*}{\hat{t}_{xd'}^*} \tag{3.19}$$

where

$$\hat{t}_{axd'} = \frac{N}{n_a} \Sigma_{s_a} z_{d'k} x_k \tag{3.20}$$

The approximate variance of $\hat{t}_{dr}$ is, according to (3.7),

$$AV(\hat{t}_{dr}) = \frac{N^2}{n_a}\left(1 - \frac{n_a}{N}\right) S_{yd}^2 + \frac{N^2}{n}\left(1 - \frac{n}{n_a}\right) E_{SI} E_{\mathbf{X}}\left[s_{Ed}^2\right] \tag{3.21}$$

where

$$S_{yd}^2 = \frac{1}{N-1} \Sigma_U \left(y_{dk} - \tilde{y}_d\right)^2 \quad ; \quad \tilde{y}_d = \frac{1}{N} \Sigma_U y_{dk} \tag{3.22}$$

$$s_{Ed}^2 = \frac{1}{n_a - 1} \Sigma_{s_a} \left(y_{dk} - \mathbf{x}'_k \mathbf{b}_{ad}\right)^2 \tag{3.23}$$

and element d' in $\mathbf{b}_{ad}$ is

$$b_{add'} = \frac{\Sigma_{s_a} z_{d'k} y_{dk}}{\Sigma_{s_a} x_{d'k}} \tag{3.24}$$

The variance of $\hat{t}_{dr}$ is estimated by

$$\hat{V}(\hat{t}_{dr}) = \frac{N^2}{n_a}\left(1 - \frac{n_a}{N}\right) s_{yd}^2 + \frac{N^2}{n}\left(1 - \frac{n}{n_a}\right) s_{ed}^2 \tag{3.25}$$

where

$$s_{yd}^2 = \frac{1}{n-1}\Sigma_s\left(y_{dk} - \hat{\bar{y}}_d\right)^2 \quad ; \quad \hat{\bar{y}}_d = \frac{1}{n}\Sigma_s y_{dk} \tag{3.26}$$

and

$$s_{ed}^2 = \frac{1}{n-1}\sum_{d'=1}^{D} g_{d'}^2 \Sigma_{s_{d'}}\left(y_{dk} - x_{d'k}b_{dd'}\right)^2 \tag{3.27}$$

The set $s_{d'}$ contains all units $k \rightarrow d'$ in the sample s and $g_{d'} = 1$ for every k or equal to $\hat{t}_{axd'} / \hat{t}^*_{xd'}$ for every $k \rightarrow d'$.

3.3 A Comparison between the *m*GREG Estimator and the π* Estimator under the *SI* Design

We can look upon the technique with two-phase sampling as an insurance when we suspect measurement and/or classification errors. The premium that we have to pay, disregarding the costs, is increased variance of the estimate. The variance *may* be decreased in two-phase sampling when an estimator that makes efficient use of the auxiliary variables is used. For example, when the parameter of interest is a function of simple totals, it is not a matter of course that the variance of the *m*GREG estimator is smaller than the variance of the π* estimator. This may be so even when the *m*GREG estimator for each total is more efficient than the π* estimator.

Recall that

$$\begin{aligned} AV(\hat{t}_{dr}) &= \frac{N^2}{n_a}\left(1 - \frac{n_a}{N}\right)S_{yd}^2 + \frac{N^2}{n}\left(1 - \frac{n}{n_a}\right)E_{SI}E_{\mathbf{X}}\left[s_{Ed}^2\right] = \\ &= V_1 + AV_2 \end{aligned} \tag{3.28}$$

From the theory of regression analysis it is known that $s_{Ed}^2 = s_{ayd}^2\left(1 - R_{ad}^2\right)$ where R_{ad}^2 is the coefficient of determination and s_{ayd}^2 is the variance of y in the sample s_a. It is also known that R_{ad}^2 is the square of the correlation between the true y-values and the predicted y-values with the error-prone x-variables as predictors in the sample s_a.

The coefficient R_{ad}^2 is called the "coefficient of reliability" by Tenenbein (1970), while 1-R_{ad}^2 is called the "index of inconsistency" by Biemer & Stokes (1991).

Let us approximate $E_{SI}E_{\mathbf{X}}\left[s_{Ed}^2\right]$ with $S_{yd}^2\left(1-R_d^2\right)$ in the population U. Then the approximate variance can be written

$$AV(\hat{t}_{dr}) \approx N^2 S_{yd}^2 \left\{ \frac{(1-R_d^2)}{n} + \frac{R_d^2}{n_a} - \frac{1}{N} \right\} \tag{3.29}$$

It is easy to see that the variance of the π^* estimator will always be at least as large as that of the *m*GREG estimator.

Remark 3.3.1. An approximate expression for (3.28) can be obtained by using the measurement model in section 1.5.

Let $E_{\mathbf{X}}\left[s_{Ed}^2\right]$ be approximated by

$$E_{\mathbf{X}}\left[s_{Ed}^2\right] \approx \frac{1}{n_a - 1}\left\{ \Sigma_{s_a} y_{dk}^2 + \Sigma_{s_a} \sum_{d'=1}^{D} \left(\frac{\Sigma_{s_a} y_{dk}\theta_{d'k}}{\Sigma_{s_a} \mu_{xk}^{(1)}\theta_{d'k}} \right)^2 \left(\sigma_{xk}^2 \theta_{d'k} + \left(\mu_{xk}^{(1)}\right)^2 \theta_{d'k} \right) - \right. \tag{3.30}$$

$$\left. -2\Sigma_{s_a} \sum_{d'=1}^{D} \frac{\Sigma_{s_a} y_{dk}\theta_{d'k}}{\Sigma_{s_a} \mu_{xk}^{(1)}\theta_{d'k}} y_{dk}\mu_{xk}^{(1)}\theta_{d'k} \right\}$$

Then an approximation for the expectation $E_{p_a}[\cdot]$ of (3.30) is obtained by substituting N for n_a and Σ_U for Σ_{s_a}.

Consider the situation described by Tenenbein (1972) where the *SI* design is used in both phases with $y_{dk}=\zeta_{dk}$, $x_{dk}=z_{dk}$, $\mu_{xk}^{(1)}=1$, $\sigma_{xk}^2=0$, $\theta_{d'k}=\theta_{d'|d}$, $k \in U_d$. Define $\pi_{d'} =$ $= \Sigma_U \theta_{d'k}/N$ and $P_d = \Sigma_U \zeta_{dk}/N$. Then (3.28) is approximated by

$$AV(\hat{t}_{dr}) \approx N^2 P_d(1-P_d)\left(\frac{1-R_d^2}{n} + \frac{R_d^2}{n_a} \right) - NP_d(1-P_d) \tag{3.31}$$

where
$$R_d^2 = \frac{P_d}{1-P_d}\left(\sum_{d'=1}^{D} \frac{\theta_{d'|d}^2}{\pi_{d'}} - 1 \right) \tag{3.32}$$

The expressions (3.31) and (3.32) are the same as in Tenenbein (1972) apart from the last term $NP_d(1\text{-}P_d)$ which comes from the finite population correction (fpc) in V_1.

The accuracy of the approximation (3.31) was confirmed in Andersson (1992), where the mean of the estimated variances from a simulation study was compared with V_1+AV_2. The approximation (3.30) has also been confirmed by a simulation study (see Appendix B). □

CHAPTER 4

Estimation of Domain Means and Relative Domain Sizes — Classification Errors but No Measurement Errors

In this chapter the problem of estimating the mean value of the response variable y in a domain and the relative size of that domain is considered. The values of the y-variable are assumed to be observed without errors for every unit $k \in s_a$ and every s_a but not all units in the sample s_a are correctly classified, this is an instance of case II in chapter 1.

Selén (1986), Mak (1988) and Mak & Li (1988) have suggested different estimation methods for the mean values in domains when $p_a(\cdot)$ is the *SIR* design and $p(\cdot \mid s_a)$ is the *SI* design. Both Mak (1988) and Mak & Li (1988) only consider case II, i.e. when $x_{dk}=z_{dk}y_k$ for every $k \in s_a$ and every s_a, while Selén (1986) also depicts the situation with misclassification and measurement errors, case IV, without giving any details about the estimator. A complicated Maximum-Likelihood (ML) estimator which demands an iterative solution that may not converge was also suggested by Selén, but it is not presented here.

Tenenbein (1970, 1971, 1972) suggested an ML estimator when the design $p_a(\cdot)$ is the *SIR* design, $p(\cdot \mid s_a)$ is the *SI* design and $y_{dk}=\zeta_{dk}$ and $x_{dk}=z_{dk}$ for every $k \in s_a$ and every s_a, i.e. the units in the samples s_a and s are classified only.

4.1 Estimation of Relative Domain Sizes

Tenenbein (1970, 1971, 1972) suggests an ML estimator in order to estimate the proportion $P_d = N_d/N$ by using a two-phase sampling scheme in the case of a misclassification model studied by Bross (1954). The misclassification model can be seen as a special case of the model m described in chapter 1. Tenenbein assumes that the classification probabilities are the same for every $k \in U_{d'}$ (d'=1, 2, ..., D), i.e. $\theta_{dk} = \theta_{d|d'} =$ $=\Pr(k \to d \mid k \in U_{d'})$.

Swensson (1973) showed that Tenenbein's estimator is equivalent to a regression estimator when D=2 and $p_a(\cdot)$ and $p(\cdot \mid s_a)$ are *SI* designs. Under the same designs the results in section 3.2 can be used to show that Tenenbein's estimator is a special case also of an estimator based on the *m*GREG estimator when $D \geq 2$.

Let $n_{ad'}$ be the number of units allocated to class d' in sample s_a where procedure F is used, let $n_{\cdot d'}$ be the number of units allocated to class d' in sample s when F is used and let $n_{dd'}$ be the number of units in the sample s that are identified as belonging to category d by procedure T and allocated to class d' by procedure F. Then element d' in $\mathbf{b}_d$ and g_k are calculated as

$$b_{dd'} = \frac{n_{dd'}}{n_{\cdot d'}} \text{ and } g_k = \frac{n_{ad'}}{n_a} \frac{n}{n_{\cdot d'}}, \quad k \rightarrow d' \tag{4.1}$$

Tenenbein's estimator and the *m*GREG estimator of P_d are then given by

$$\hat{P}_d = \frac{\hat{N}_{dr}}{N} = \frac{1}{n_a} \sum_{d'=1}^{D} n_{ad'} \frac{n_{dd'}}{n_{\cdot d'}} \tag{4.2}$$

It was shown in Andersson (1992) that the approximate variance of (4.2) is nearly the same as the variance of Tenenbein's estimator when the expression (3.21) in section 3.2 is used. It was also shown that the variance estimators are similar when the estimator (3.25) is used in combination with a suitable choice of g_k.

4.2 Estimation of Domain Means

The mean value of the y-variable in domain $d = 1, \dots D$ is defined as

$$\bar{y}_d = t_d / N_d \tag{4.3}$$

When N_d is unknown, the parameter $\bar{y}_d$ is a ratio of two unknowns, each of which has to be estimated in order to get an estimate of the mean value. It was shown in chapter 1 that the estimator of $\bar{y}_d$, based on the error-prone values in the sample s_a, will generally be biased when the units are not correctly classified.

Different estimators have been suggested when a two-phase sampling scheme is used in the case of a misclassification and the design $p_a(\cdot)$ is *SIR* and $p(\cdot \mid s_a)$ is *SI*. However, it is easy to construct an estimator based on the *m*GREG estimator for any parameter that is a rational function of totals when arbitrary designs are used for the first- and second-phase samples (see Appendix A).

4.2.1 An *m*GREG-Based Estimator of the Domain Mean

The *m*GREG estimator is an approximately unbiased estimator of a total when two-phase sampling is used. It seems natural to use the *m*GREG estimators of t_d and N_d in (4.3) to obtain an *m*GREG-based estimator of $\bar{y}_d$

$$\hat{\bar{y}}_{dr} = \hat{t}_{dr} / \hat{N}_{dr} \quad (4.4)$$

where $\hat{t}_{dr}$ is defined as in (3.2) and $\hat{N}_{dr}$ is defined in a similar way with $y_{dk}=\zeta_{dk}$.

The variance of $\hat{\bar{y}}_{dr}$ is obtained by using the first-order Taylor series approximation of the function of estimated totals around their expected values. The variance of $\hat{\bar{y}}_{dr}$ is estimated by replacing the population totals by their sample estimates in the derivatives. Then we get

$$\hat{V}(\hat{\bar{y}}_{dr}) = \frac{1}{\hat{N}_{dr}^2}\left\{\hat{V}(\hat{t}_{dr}) + \hat{\bar{y}}_{dr}^2\hat{V}(\hat{N}_{dr}) - 2\hat{\bar{y}}_{dr}\hat{C}(\hat{t}_{dr},\hat{N}_{dr})\right\} \quad (4.5)$$

where $\hat{C}(\hat{t}_{dr},\hat{N}_{dr})$ is obtained as

$$\hat{C}(\hat{t}_{dr},\hat{N}_{dr}) = \sum\sum_s \frac{\Delta_{akl}}{\pi_{kl}^*}\frac{y_{dk}}{\pi_{ak}}\frac{\zeta_{dl}}{\pi_{al}} + \sum\sum_s \frac{\Delta_{kl|a}}{\pi_{kl|a}}\frac{g_{yk}e_{ydk}}{\pi_k^*}\frac{g_{zl}e_{zdl}}{\pi_l^*} \quad (4.6)$$

where $e_{ydk} = y_{dk} - \mathbf{x}'_{dk}\mathbf{b}_{yd}$, $e_{zdk} = \zeta_{dk} - \mathbf{z}'_{dk}\mathbf{b}_{zd}$, g_{yk} is defined by (2.21), g_{zk} is defined by (2.21) with $\mathbf{x}_k=\mathbf{z}_k$, $\mathbf{b}_{yd}$ is defined by (2.13) and $\mathbf{b}_{zd}$ is defined by (2.13) with $y_{dk}=\zeta_{dk}$ and $\mathbf{x}_k=\mathbf{z}_k$. The variance $\hat{V}(\hat{t}_{dr})$ is obtained from (3.8), while $\hat{V}(\hat{N}_{dr})$is obtained from (3.8) with $y_{dk}=\zeta_{dk}$ and $e_{dk}=e_{zdk}$.

4.2.2 Selén's and Mak's Estimators

Let $\mathbf{\Phi}$ be a matrix with element *dd'* equal to the *a posteriori* probability $\phi_{d|d'}=\Pr(k\in U_d | k\rightarrow d')$, $d=1, \ldots, D$, $d'=1, \ldots, D$. Define $\hat{\bar{\mathbf{x}}}_a = (\hat{\bar{x}}_{a1},\ldots,\hat{\bar{x}}_{ad},\ldots,\hat{\bar{x}}_{aD})'$ as the mean values of the observed *x*-values from the units $k\in s_a$ in the different domains and let $\bar{\mathbf{y}} = (\bar{y}_1,\ldots,\bar{y}_d,\ldots,\bar{y}_D)'$ be the mean values of *y* in the different domains. Then $E(\hat{\bar{\mathbf{x}}}_a) = \mathbf{\Phi}'\bar{\mathbf{y}}$.

If there exists an inverse $\mathbf{\Phi}^-$ of $\mathbf{\Phi}'$ we can use $\hat{\bar{\mathbf{y}}} = \hat{\mathbf{\Phi}}^-\hat{\bar{\mathbf{x}}}_a$ as an estimator of $\bar{\mathbf{y}}$. Selén and Mak use different approaches to estimate the elements in $\mathbf{\Phi}$; however, in both cases the estimator can be written on the form

$$\hat{\hat{y}}_d^* = w_d \sum_{d'=1}^{D} \hat{\phi}^{dd'} \hat{\bar{x}}_{ad'}^{(-)} + (1 - w_d)\hat{\bar{y}}_d \tag{4.7}$$

where $\hat{\phi}^{dd'}$ is the estimated element dd' in $\hat{\mathbf{\Phi}}^{-}$ and

$$\hat{\bar{x}}_{ad'}^{(-)} = \frac{1}{n_{ad'} - n_{\cdot d'}} \sum_{(s_a - s)} x_{d'k} \tag{4.8}$$

is the mean value of the observed $x_{d'k}$-values, based on the units in $(s_a–s)$, i.e. the units that do not belong to the second-phase sample s. The mean value of the unobserved y_{dk}-values, $k \in (s_a–s)$, in (4.7) are predicted as

$$\sum_{d'=1}^{D} \hat{\phi}^{dd'} \hat{\bar{x}}_{ad'}^{(-)} \tag{4.9}$$

The weight w_d may be chosen as $w_d=(n_a–n)/n_a$, as suggested by Selén (1986), or as a function of the variances of the two components in the estimator, as suggested by Mak (1988). Selén and Mak use different approaches to estimate the elements in $\mathbf{\Phi}$.

Mak (1988) estimates the elements in $\mathbf{\Phi}$ with $\hat{\phi}_{d|d'} = n_{dd'}/n_{\cdot d'}$, while Selén (1986) uses the fact that $\mathbf{\Phi}$ can be written as

$$\mathbf{\Phi} = diag(\mathbf{P})\mathbf{\Theta} diag(\mathbf{\Theta}'\mathbf{P})^{-1} \tag{4.10}$$

where $diag(\cdot)$ is a diagonal matrix, $\mathbf{\Theta}$ is a matrix with element dd' equal to $\theta_{d'|d}=\Pr(k \to d' \mid k \in U_d)$ which is estimated by $\hat{\theta}_{d'|d} = n_{dd'} / n_{d\cdot}$ and $\mathbf{P}$ is a vector with element d equal to the relative size of domain d which is estimated by

$$\hat{P}_d = \frac{1}{n_a}\left(n_{d\cdot} + \sum_{d'=1}^{D} \hat{\theta}^{dd'} (n_{ad'} - n_{\cdot d'}) \right) \tag{4.11}$$

where $\hat{\theta}^{dd'}$ is element dd' in the inverse $\hat{\mathbf{\Theta}}^{-}$ of $\hat{\mathbf{\Theta}}'$.

Mak (1988) demonstrates in a simulation study that his estimator is more efficient then Selén's, at least in the cases he studied.

The simulation studies reported in Selén (1986) and Mak (1988) show that both estimators occasionally give extreme estimates due to the inversion of the matrix $\hat{\mathbf{\Phi}}$, which may be ill-conditioned for some samples. These extreme estimates seem to occur more frequently when Selén's estimator is used.

4.2.3 Mak & Li's Estimator

Mak & Li motivate their estimator in the following way. The mean $\hat{\bar{x}}_d$ computed from the second-phase sample s should be correlated with $\hat{\bar{y}}_d$, furthermore the means

$\hat{\bar{x}}_{d'}$, $d' \neq d$, should also be correlated with $\hat{\bar{y}}_d$ through the presence of misclassified units. Then, under suitable regularity conditions, the distribution of $(\hat{\bar{y}}_d, \hat{\bar{x}}_1, \dots, \hat{\bar{x}}_D)'$ is multivariate normal asymptotically as n becomes large. Applying the restricted maximum likelihood estimation approach (Cox & Hinkley 1974) to the asymptotic likelihood of $(\hat{\bar{y}}_d, \hat{\bar{x}}_1, \dots, \hat{\bar{x}}_D)'$, Mak & Li obtain the estimator

$$\hat{\bar{y}}_{MLd} = \hat{\bar{y}}_d + (\hat{\bar{\mathbf{x}}}_a - \hat{\bar{\mathbf{x}}})' \mathbf{b}_d \tag{4.12}$$

By Theorem 1 in Mak & Li (1988) the coefficient $\mathbf{b}_d$ is computed as $\mathbf{b}_d = \hat{\mathbf{u}}'_d \hat{\mathbf{V}}^{-1}$, where $\hat{\mathbf{V}} = diag(\hat{v}_{11}, \dots, \hat{v}_{d'd'}, \dots, \hat{v}_{DD})$, $\hat{\mathbf{u}}_d = (\hat{u}_{d1}, \dots, \hat{u}_{dd'}, \dots, \hat{u}_{dD})'$ and $diag(\cdot)$ denotes a diagonal matrix.

Mak & Li's expressions for $\hat{u}_{dd'}$ and $\hat{v}_{d'd'}$, using my notation, are

$$\hat{u}_{dd'} = \frac{n_{dd'} \times n}{n_{d\cdot} \times n_{\cdot d'}} \left(\frac{1}{n_{dd'}} \Sigma_s y_{dk} x_{d'k} - \hat{\bar{y}}_d \hat{\bar{x}}_{dd'} - \hat{\bar{y}}_{dd'} \hat{\bar{x}}_{d'} + \hat{\bar{y}}_d \hat{\bar{x}}_{d'} \right) \tag{4.13}$$

and

$$\hat{v}_{d'd'} = \frac{n}{n_{\cdot d'}} \left(\frac{1}{n_{\cdot d'}} \Sigma_s x_{d'k}^2 - \hat{\bar{x}}_{d'}^2 \right) \tag{4.14}$$

The variance estimator proposed by Mak & Li is

$$\hat{V}\left(\hat{\bar{y}}_{MLd}\right) = \frac{1}{n}\left(w - (1 - n / n_a) \hat{\mathbf{u}}'_d \hat{\mathbf{V}}^{-1} \hat{\mathbf{u}}_d\right) \tag{4.15}$$

where $w = \frac{n}{n_{d\cdot}}\left(\Sigma_s y_{dk}^2 / n_{d\cdot} - \hat{\bar{y}}_d^2\right)$ and $\hat{\mathbf{u}}_d$ and $\hat{\mathbf{V}}$ are calculated according to (4.13) and (4.14).

4.3 A Simulation Study — A Comparison Between Four Estimators of Domain Means

A simulation study was carried out to compare the sampling distribution properties of the different estimators for domain means with respect to bias and variance and other distribution properties. The estimators studied are the *m*GREG-based estimator of the mean and the estimators suggested by Selén (1986), Mak (1988) and Mak & Li (1988). The simulation study was done using artificial data and the programming was done by the use of APL*PLUS 7.0, STSC Inc. and an IBM PS/2 model 70 computer.

4.3.1 The Artificial Population

A population U of 5,000 units was generated. The population was divided into two disjoint domains U_1 and U_2 with sizes N_1=1,000 and N_2=4,000, i.e. P_1=0.2 and P_2=0.8. The values of three response variables y_1, y_2, y_3 were generated within each domain in the following way.

Variable y_1 is dichotomous with two possible values, 0 and 1, with p=0.8 in U_1 and p=0.4 in U_2, i.e. the mean values of y_1 are $\bar{y}_{11}$=0.8 in U_1 and $\bar{y}_{12}$=0.4 in U_2. The variances are S_{11}^2=0.16 in U_1 and S_{12}^2=0.24 in U_2. The values of y_2 were generated from a normal distribution within each domain and with the same means and variances as for y_1. The values of y_3 were generated from a gamma distribution to get skew distributions and the generated values were transformed to get the same means and variances as y_1 and y_2.

4.3.2 The Simulation Study

The simulation study was carried out by repeating the sampling procedure a number of times. In each repetition, a first-phase *SI* sample s_a of size n_a=500 was drawn and a second-phase *SI* sample s of size n=100 was drawn from s_a. Each unit in s_a was classified into class 1 or class 2 by using two different sets of classification probabilities, Cases A) and B), while the units in s were classified correctly.

Case A)
$$\begin{cases} \theta_{1|1}^{A} = 0.9 \\ \theta_{2|2}^{A} = 0.8 \end{cases} \tag{4.16}$$

Case B)
$$\begin{cases} \theta_{1|1}^{B1} = 0.92 & \text{if } (y_{1k}=1)\&(k \in U_1) \\ \theta_{2|2}^{B1} = 0.89 & \text{if } (y_{1k}=1)\&(k \in U_2) \end{cases} \qquad \begin{cases} \theta_{1|1}^{B0} = 0.82 & \text{if } (y_{1k}=0)\&(k \in U_1) \\ \theta_{2|2}^{B0} = 0.74 & \text{if } (y_{1k}=0)\&(k \in U_2) \end{cases} \tag{4.17}$$

Case B) was chosen to study the effects on Selén's and Mak's estimators of non-zero correlation between θ_{dk} and the y-values. The misclassification probability was chosen to satisfy the relation

$$\theta_{d'|d}^{A} = \theta_{d'|d}^{B1}\bar{y}_{1d} + \theta_{d'|d}^{B0}(1-\bar{y}_{1d}) \tag{4.18}$$

which means that $\Sigma_{U_1}\theta_{1k}^{A} = \Sigma_{U_1}\theta_{1k}^{B}$ and $\Sigma_{U_2}\theta_{2k}^{A} = \Sigma_{U_2}\theta_{2k}^{B}$, i.e. the expected numbers of units allocated to domains 1 and 2 are the same in Cases A) and B).

The values of y_2 and y_3 were sorted by y_1, i.e. y_1=1 for the smallest values of y_2 and y_3, which means that θ_{dk} is correlated with y_1, y_2 and y_3 in Case B).

The division of the population into one small and one large domain was chosen to conform with Selén's (1986) and Mak & Li's (1988) simulation studies. The sample sizes in the first and second phases and the y_1 and y_2 variables were chosen for the same reason, while y_3 was chosen to conform with one of the variables studied by Mak (1988).

The misclassification Case A) is the same as one of the studied cases in Selén and Mak & Li, while the idea of Case B) comes from Mak & Li.

The population size was set at 5,000, which is large enough to make the fpc in the first phase small but not too large to burden the computations.

In each Case, A) and B), 1,000 samples were taken. It is believed that 1,000 samples is enough to get reliable estimates of the bias in $\hat{\bar{y}}_{idj}$, but the number of samples may not be sufficient to get reliable estimates of the variances of the estimated means.

For each sample *j*, (*j*=1, ..., 1,000) the estimates $\hat{\bar{y}}_{idj}$ (*i*=1, 2, 3, *d*=1, 2) for variable *i* in domain *d* were calculated for each proposed estimator. Furthermore, for the Mak & Li estimator and the *m*GREG-based estimator the estimated variances $\hat{V}(\hat{\bar{y}}_{idj})$ and the intervals $\hat{\bar{y}}_{idj} \pm 1.96\sqrt{\hat{V}(\hat{\bar{y}}_{idj})}$ were calculated and it was checked whether the interval covered the true mean $\bar{y}_{id}$.

In each first-phase sample s_a, the unadjusted estimate of the mean, $\hat{\bar{x}}_{ad}$, based on the error-prone classification, was computed. For these estimates, small variances, due to the large sample size and large bias, were expected.

It was also expected that the estimates of the means in domain 1 would be more sensitive to misclassification than those of the means in U_2, because U_1 is smaller than U_2.

4.3.3 Results of the Simulation Study

Bias

The difference between the mean of the estimated mean values and the true mean value in a domain is an estimate of the bias in the estimator of $\bar{y}_{id}$.

The hypothesis that the bias $E\left[\hat{\bar{y}}_{id}\right]-\bar{y}_{id}=0$ for each i and d can be tested by the variable

$$z=\frac{\bar{\hat{\bar{y}}}_{id}-\bar{y}_{id}}{\sqrt{S^2\left(\hat{\bar{y}}_{id}\right)/1{,}000}} \tag{4.19}$$

where $$S^2\left(\hat{\bar{y}}_{id}\right)=\frac{1}{999}\sum_{j=1}^{1{,}000}\left(\hat{\bar{y}}_{idj}-\bar{\hat{\bar{y}}}_{id}\right)^2 \quad ; \quad \bar{\hat{\bar{y}}}_{id}=\frac{1}{1{,}000}\sum_{j=1}^{1{,}000}\hat{\bar{y}}_{idj} \tag{4.20}$$

The test variable z in (4.19) is approximately distributed as $N(0, 1)$ when $E\left[\hat{\bar{y}}_{id}\right]=\bar{y}_{id}$. The test is used as an indicator of bias in the estimators. The results of the tests are given in table 4.1.

NOTE: Selén's, Mak's and the *m*GREG estimators gave estimates of $\bar{y}_{11}$ and $\bar{y}_{12}$ that fell outside (0, 1), in some samples. These estimates were called inadmissible by Selén (1986). When an estimate larger than 1 was obtained it was set at 1; no negative estimates could be obtained in this study. In Case A) the proportions of inadmissible estimates were 9.6% for Selén's estimator, 3.9% for Mak's and 2.5% for the *m*GREG estimator, and in Case B) the proportions were 1.5%, 0.1% and 4.5% respectively. □

The observed bias is highly significant when the unadjusted estimator is used, as was expected. In Case A), which is the case where the assumptions for Selén's and Mak's estimators are met, the observed bias for Selén's estimator is significant at the 0.1% level in domain 1, while the observed biases are negligible when the other estimators are used.

In Case B), where there are correlations between the *y*-variable and the classification probability, the estimated biases were highly significant in domains 1 and 2 when Selén's and Mak's estimators were used. In domain 2, the observed biases for Selén's and Mak's estimators were even larger than the biases for the unadjusted estimator. Mak & Li's estimator shows medium and low significances in Cases A) and B) although the observed values of the biases are small. The *m*GREG estimator shows only one, low, significant value.

Although unbiasedness is a desirable property it may be reasonable to use an estimator that is only approximately unbiased since it may, for example, have highly desirable variance properties. However, it is generally agreed that estimators with large bias should be avoided. Hajek (1971) expresses this rule resolutely: "...greatly biased estimates are poor no matter what other properties they have".

Even if the bias is significant it may not be *relevant*. When the number of samples in a Monte Carlo study is large enough the bias will be significantly different from zero when an approximately unbiased estimator is used in (4.19). Thus we need a criterion to decide whether a bias is relevant or not. The absolute bias ratio, *ABR*, is such a criterion:

$$ABR = \frac{\left|E\left[\hat{\bar{y}}_{id}\right] - \bar{y}_{id}\right|}{\sqrt{V(\hat{\bar{y}}_{id})}} \tag{4.21}$$

If $\hat{\bar{y}}_{id}$ is normally distributed, the coverage probability of a calculated 95% confidence interval will be distorted when *ABR* is large. It can be shown that the distortion effect is not very large when *ABR*<0.5. When *ABR*=0.5 the coverage probability is 92% but when *ABR*=1.0 the coverage probability is only 83%. In the simulation study reported here with 1,000 samples, the observed values of *ABR* are given in table 4.2.

Besides the unadjusted estimator, Selén's estimator shows large biases as well as large *ABR*s in Cases A) and B), while Mak's estimator fails on both criteria in Case B). Due to the large bias for Selén's estimator it will be excluded from further investigations. In practice little will be known about the correlation between θ_{dk}s and y_k, and thus Mak's estimator will be excluded from further investigations too.

When the test for bias and the *ABR* criteria are used to select proper estimators, only Mak & Li's and the *m*GREG estimators remain. The distributions of 1,000 estimates $\hat{\bar{y}}_{id}$ and $\hat{V}(\hat{\bar{y}}_{id})$, i=1, 2, 3, d=1, 2, Cases A) and B) for these two estimators are shown in figures 4.1-4.48.

Since the observed biases are small, it is of interest to study the variances of their estimates, to see if the proposed variance estimators are at least approximately unbiased and if the estimated confidence intervals have the intended coverage rates.

Comparison of Variance Estimators and Coverage Rates

The variances of the estimated values were calculated according to (4.5) and (4.15) for Mak & Li's and the *m*GREG estimators, respectively.

The estimated means can be considered at least approximately normally distributed, see figures 4.1-4.64, thus $S^2(\hat{\bar{y}}_{id})$ is distributed approximately as $c\chi^2$ with 999 degrees of freedom, where c is a constant. Then it is possible to calculate the approximate upper (U) and lower (L) limits for the variance $V(\hat{\bar{y}}_{id})$ as

$$S_L^2\left(\hat{\bar{y}}_{id}\right)=\frac{999\cdot S^2\left(\hat{\bar{y}}_{id}\right)}{\chi^2_{0.975}(999)} \quad ; \quad S_U^2\left(\hat{\bar{y}}_{id}\right)=\frac{999\cdot S^2\left(\hat{\bar{y}}_{id}\right)}{\chi^2_{0.025}(999)} \tag{4.22}$$

where $\chi^2_{0.975}(999)=1{,}088.5$ and $\chi^2_{0.025}(999)=913.3$.

An approximate 95% confidence interval for the mean of the estimated variances is calculated as

$$\bar{\hat{V}}\left(\hat{\bar{y}}_{id}\right)\pm 1.96\sqrt{S^2\left(\hat{V}(\hat{\bar{y}}_{id})\right)} \tag{4.23}$$

since the distributions of $\hat{V}(\cdot)$ are not extremely skew, (see figures 4.1-4.64).

The upper and lower limits (4.22) and the confidence intervals (4.23) are given in table 4.2 for Mak & Li's estimator and in table 4.3 for the *m*GREG estimator.

Now, at least two questions arise: 1) are the proposed variance estimators for Mak & Li's and the *m*GREG estimators approximately unbiased, and 2) are the variances of the estimated mean values approximately equal for Mak & Li's and the *m*GREG estimators?

The variance estimator $\hat{V}(\hat{\bar{y}}_{id})$ is considered approximately unbiased for the variance $V(\hat{\bar{y}}_{id})$ if the observed difference $\bar{\hat{V}}(\hat{\bar{y}}_{id})-S^2(\hat{\bar{y}}_{id})$ does not differ significantly from zero. However, it is difficult or maybe even impossible to obtain the variance of $\bar{\hat{V}}(\hat{\bar{y}}_{id})-S^2(\hat{\bar{y}}_{id})$ analytically. One possible solution is to compute the jackknife estimate of the difference and its variance.

By excluding 40 samples from 1,000 at a time we get 25 pseudovalues of the difference. The variance is estimated from the 25 pseudovalues and the difference is tested by using a *t*-test with 24 degrees of freedom. The results of the tests are shown in tables 4.3, 4.4 and 4.6.

Most of the jackknifed differences of Mak & Li's estimator are significant at the 0.1% level or at the 1% level. This indicates a bias in Mak & Li's variance estimator, at least for the cases in this study.

When the *m*GREG estimator is studied, the jackknifed differences are significant at the 0.1% level for $\bar{y}_{11}$ in Case B), probably due to the inadmissible estimates. The differences are significant at the 5% level for $\bar{y}_{11}$ in Case A) and for $\bar{y}_{22}$ and $\bar{y}_{32}$ in Cases A) and B). Thus the bias of the variance estimator for the *m*GREG estimator is less serious than the bias of Mak & Li's variance estimator in most of the studied cases.

The observed variances $S^2(\hat{\bar{y}}_{id})$ are usually smaller when Mak & Li's estimator is used than when the *m*GREG estimator is used. The jackknife technique is used to test whether the $S^2(\hat{\bar{y}}_{id})$ differs significantly for Mak & Li's and the *m*GREG estimators. The jackknifed differences are significant at the 0.1% level in domain 1 for all three variables and in the two Cases. In domain 2 the differences are significant at the 0.1% and 1% levels in Case A) while they are significant only at the 5% level for variable y_1 in Case B).

Other criteria of interest regarding an estimator are the coefficient of variation, *cve*, of the estimator and the coverage rate of the confidence intervals constructed by using the estimated variances. The coefficient of variation is calculated as

$$cve\left(\hat{\bar{y}}_{id}\right) = \frac{\sqrt{S^2(\hat{\bar{y}}_{id})}}{\hat{\bar{y}}_{id}} \tag{4.24}$$

and a confidence interval in sample *j* is calculated as $\hat{\bar{y}}_{idj} \pm 1.96\sqrt{\hat{V}(\hat{\bar{y}}_{idj})}$. The coverage rate, *CI*95, of the 1,000 intervals was calculated for each estimator as the proportion of intervals covering the true mean value.

The hypothesis that the coverage rate of the estimated intervals is 0.95, is tested by

$$z = \frac{\hat{p} - 0.95}{\sqrt{0.95 \times 0.05/1{,}000}} \tag{4.25}$$

where $\hat{p}$ is the observed coverage rate. The test variable z is approximately distributed as $N(0, 1)$ if the true coverage rate is 0.95. The *cve*s and the coverage rates are given in table 4.5.

The *cve*s are about the same for the estimators. The estimated intervals in domain 1 are too short when Mak & Li's estimator is used because the differences between the observed and intended coverage rates are significant at the 0.1% level. When the *m*GREG estimator is used there are two significant intervals at the 5% level.

The Problem of Inadmissible Estimates

When a inadmissible estimate was obtained in the simulation study, the estimate was replaced by the value 1. This does not conform with how Selén, Mak and Mak & Li treated these cases. They excluded the samples that gave the inadmissible estimates from the computations of bias and variance. In a real-life situation this means that whenever the estimate of $\bar{y}_{1d}$ is outside the range (0, 1) a new sample is taken, which is an unrealistic approach.

The inadmissible estimates obtained by Selén's and Mak's estimators are ascribable to the fact that a matrix has to be inverted in the calculation of the estimate, and for some samples this matrix might be ill-conditioned.

When the *m*GREG based estimator is used, the mean is estimated as the ratio between two estimated totals. The totals are estimated using different auxiliary variables in the numerator and denominator and there is no simple way to control them to get estimates between 0 and 1.

To avoid the inadmissible estimates for the *m*GREG estimator, the same auxiliary variables may be used in numerator and denominator. There are two alternatives, the x_{dk}-values and the z_{dk}-values, and it is not an easy task to judge which one to choose in order to get the smallest variance. There are several variances and covariances involved in the expression for the variance of the estimated mean.

A simulation study was carried out to investigate the sampling properties of the two possible auxiliary variables when $\bar{y}_{11}$and $\bar{y}_{12}$ were estimated by the *m*GREG estimator in Cases A) and B). The results of that study are shown in table 4.6 and in figures 4.49-4.64.

When the same auxiliary variable was used in the estimation of the numerator and denominator, no inadmissible estimates were observed. The bias is significant in one of the studied cases. The differences $\overline{\hat{V}}(\hat{\bar{y}}_{id}) - S^2(\hat{\bar{y}}_{id})$ are significant at the 5% level in domain 1 and at the 0.1% level in domain 2. Due to the decreased values of the estimated variances $\hat{V}$ in domain 1 the observed coverage rates of the estimated confidence intervals are far below the intended 95%, while they are still non-significant in domain 2. It does not seem to matter whether x or z is used as the auxiliary variable in the studied cases, the variances and the coverage rates are about the same.

4.3.4 Conclusions from the Simulation Study

On the basis of the results of this simulation study the following conclusions can be drawn:

i) Bias

In Case A) where the assumptions for Selén's and Mak's estimators are met, Selén's estimator seems to give biased estimates, at least in some cases. In Case B) where the non-correlation assumption is violated, Selén's and Mak's estimators seems to be give biased estimates, as one could expect. Thus both Selén's and Mak's estimators are considered inferior due to large bias.

Mak & Li's estimator and the *m*GREG estimator have negligible bias in all studied cases.

ii) Variance

The observed variances of Mak & Li's estimates are in general smaller than the observed variances of the *m*GREG estimates, but the variance estimator proposed by Mak & Li seems to systematically underestimate the variances in domain 1 and overestimate the variances in domain 2. The bias in the variance estimator proposed for *m*GREG is negligible in most of the studied cases, unless the same auxiliary variable is used in the numerator and the denominator.

In domain 1, the distributions of the *m*GREG-estimates of the variances are more spread out compared to those of Mak & Li. The opposite is at hand in domain 2 as can be seen from figures 4.1-4.48.

iii) Distributional properties

The coefficients of variation, *cve*, are about the same when Mak & Li's and the *m*GREG estimators are used. When Mak & Li's estimator is used, the coverage rate of the calculated confidence intervals differs from the intended 95% in domain 1 due to the underestimates of the variances. The observed coverage rates are then far below 95%. When the *m*GREG estimator is used, the estimated intervals have approximately the intended coverage rates.

The conclusions from *i)-iii)* are that Selén's and Mak's estimators are considered inferior due to large bias, while Mak & Li's approach is considered inferior due to a biased variance estimator giving improper confidence intervals in many cases, whereas the *m*GREG estimator behaves well with regard to the criteria used here.

Table 4.1 *Observed bias in the estimated mean for each of the proposed estimators.*

		Estimator				
Case	*Parameter*	*Unadjusted*	*Selén*	*Mak*	*Mak & Li*	*mGREG*
A)	$\bar{y}_{11}$	-0.187^{***}	0.012^{***}	0.006^{*}	0.006^{*}	0.003
	$\bar{y}_{12}$	0.011^{***}	-0.002^{*}	-0.002	-0.002^{*}	-0.002
	$\bar{y}_{21}$	-0.189^{***}	-0.042^{**}	0.007^{*}	-0.004	0.006
	$\bar{y}_{22}$	0.012^{***}	0.001	0.001	0.003^{**}	0.001
	$\bar{y}_{31}$	-0.190^{***}	0.042^{**}	0.007^{*}	-0.004	0.006^{*}
	$\bar{y}_{32}$	0.013^{***}	0.001	0.002	0.002^{*}	0.001
B)	$\bar{y}_{11}$	-0.262^{***}	-0.132^{***}	-0.142^{***}	0.008^{**}	0.005
	$\bar{y}_{12}$	0.049^{***}	0.035^{***}	0.035^{***}	-0.002	-0.001
	$\bar{y}_{21}$	-0.130^{***}	0.166^{***}	0.125^{***}	-0.004	0.001
	$\bar{y}_{22}$	-0.018^{***}	-0.028^{***}	-0.028^{***}	0.002^{**}	0.001
	$\bar{y}_{31}$	-0.135^{***}	0.154^{***}	0.114^{***}	-0.004	0.001
	$\bar{y}_{32}$	-0.015^{***}	-0.025^{***}	-0.025^{***}	0.002	0.001

Test of bias
*** significant at 0.1% level, ** significant at 1% level, * significant at 5% level

Table 4.2 *Observed ABRs of the estimated mean for each of the proposed estimators*

		Estimator				
Case	*Parameter*	*Unadjusted*	*Selén*	*Mak*	*Mak & Li*	*mGREG*
A)	$\bar{y}_{11}$	5.32	0.117	0.074	0.073	0.032
	$\bar{y}_{12}$	0.401	0.072	0.059	0.063	0.041
	$\bar{y}_{21}$	5.09	0.089	0.073	0.046	0.060
	$\bar{y}_{22}$	0.460	0.038	0.053	0.094	0.037
	$\bar{y}_{31}$	5.08	0.090	0.075	0.047	0.062
	$\bar{y}_{32}$	0.520	0.038	0.055	0.073	0.039
B)	$\bar{y}_{11}$	7.39	1.37	1.82	0.096	0.050
	$\bar{y}_{12}$	1.83	1.34	1.35	0.061	0.032
	$\bar{y}_{21}$	4.11	0.627	1.15	0.049	0.013
	$\bar{y}_{22}$	0.690	0.945	0.950	0.083	0.037
	$\bar{y}_{31}$	3.78	0.595	1.06	0.047	0.017
	$\bar{y}_{32}$	0.590	0.861	0.863	0.058	0.040

Table 4.3 *Approximate lower and upper 95% confidence limits for the variance of the estimated means and approximate confidence limits for the mean of the estimated variances when Mak & Li's estimator is used (all figures* $\times 10^3$).

		Estimator: Mak & Li				
Case	*Parameter*	$S_L^2(\hat{\bar{y}}_{id})$	$S^2(\hat{\bar{y}}_{id})$	$S_U^2(\hat{\bar{y}}_{id})$	$\bar{\hat{V}}(\hat{\bar{y}}_{id})$	$\pm 1.96\sqrt{S^2(\bar{\hat{V}})}$
A)	$\bar{y}_{11}$	6.34	6.90^{+++}	7.56	$5.33^{\times\times\times}$	0.149
	$\bar{y}_{12}$	0.846	0.921^{++}	1.01	$1.06^{\times\times}$	0.009
	$\bar{y}_{21}$	6.03	6.56^{+++}	7.18	$5.37^{\times\times}$	0.138
	$\bar{y}_{22}$	0.750	0.816^{+++}	0.893	$1.03^{\times\times\times}$	0.017
	$\bar{y}_{31}$	5.96	6.49^{+++}	7.11	$5.37^{\times\times}$	0.161
	$\bar{y}_{32}$	0.751	0.817^{++}	0.895	$1.02^{\times\times\times}$	0.022
B)	$\bar{y}_{11}$	6.32	6.88^{+++}	7.53	$5.57^{\times\times\times}$	0.149
	$\bar{y}_{12}$	0.805	0.876^{+}	0.959	$0.981^{\times}$	0.009
	$\bar{y}_{21}$	5.59	6.09^{+++}	6.67	$5.12^{\times\times}$	0.137
	$\bar{y}_{22}$	0.817	0.889	0.973	$1.01^{\times\times}$	0.017
	$\bar{y}_{31}$	5.70	6.20^{+++}	6.79	$5.27^{\times\times}$	0.163
	$\bar{y}_{32}$	0.826	0.899	0.984	$1.03^{\times\times}$	0.023

Test of $\bar{\hat{V}}(\hat{\bar{y}}_{id}) - S^2(\hat{\bar{y}}_{id})$

$\times\times\times$ significant at 0.1% level, $\times\times$ significant at 1% level, $\times$ significant at 5% level

Test of $S^2_{m\text{GREG}}$-$S^2_{\text{M\&L}}$

+++ significant at 0.1% level, ++ significant at 1% level, + significant at 5% level

Table 4.4 *Approximate lower and upper 95% confidence limits for the variance of the estimated means and approximate confidence limits for the mean of the estimated variances when the mGREG estimator is used (all figures* $\times 10^3$).

		Estimator: mGREG				
Case	*Parameter*	$S_L^2(\hat{\bar{y}}_{id})$	$S^2(\hat{\bar{y}}_{id})$	$S_U^2(\hat{\bar{y}}_{id})$	$\bar{\hat{V}}(\hat{\bar{y}}_{id})$	$\pm 1.96\sqrt{S^2(\bar{\hat{V}})}$
A)	$\bar{y}_{11}$	7.42	8.08^{+++}	8.84	$8.91^{\times}$	0.298
	$\bar{y}_{12}$	0.882	0.960^{++}	1.05	0.964	0.007
	$\bar{y}_{21}$	8.21	8.94^{+++}	9.79	9.13	0.356
	$\bar{y}_{22}$	0.806	0.877^{+++}	0.960	$0.968^{\times}$	0.014
	$\bar{y}_{31}$	7.94	8.64^{+++}	9.46	8.61	0.281
	$\bar{y}_{32}$	0.790	0.861^{++}	0.943	$0.941^{\times}$	0.013
B)	$\bar{y}_{11}$	8.88	9.67^{+++}	10.6	$11.6^{\times\times\times}$	0.382
	$\bar{y}_{12}$	0.760	0.827^{+}	0.906	$0.916^{\times}$	0.006
	$\bar{y}_{21}$	6.32	6.88^{+++}	7.53	6.77	0.222
	$\bar{y}_{22}$	0.802	0.873	0.956	$0.967^{\times}$	0.013
	$\bar{y}_{31}$	6.52	7.10^{+++}	7.77	6.90	0.207
	$\bar{y}_{32}$	0.810	0.882	0.966	$0.969^{\times}$	0.014

Test of $\bar{\hat{V}}(\hat{\bar{y}}_{id}) - S^2(\hat{\bar{y}}_{id})$

$\times\times\times$ significant at 0.1% level, $\times\times$ significant at 1% level, $\times$ significant at 5% level

Test of $S^2{}_{m\text{GREG}}$-$S^2{}_{\text{M\&L}}$

+++ significant at 0.1% level, ++ significant at 1% level, + significant at 5% level

Table 4.5 *Observed coverage rates and coefficients of variation when Mak & Li's and the mGREG estimators are used.*

		Estimator			
		Mak & Li		*mGREG*	
Case	*Parameter*	*CI95 %*	*cve %*	*CI95 %*	*cve %*
A)	$\bar{y}_{11}$	86.1***	10.4	95.6	10.7
	$\bar{y}_{12}$	95.7	7.59	94.9	7.75
	$\bar{y}_{21}$	90.3***	10.1	93.4*	11.8
	$\bar{y}_{22}$	96.2	7.14	95.7	7.40
	$\bar{y}_{31}$	90.0***	10.1	93.7	11.6
	$\bar{y}_{32}$	95.5	7.15	95.7	7.34
B)	$\bar{y}_{11}$	86.3***	10.4	95.1	11.4
	$\bar{y}_{12}$	95.5	7.40	96.4*	7.19
	$\bar{y}_{21}$	91.5***	9.75	94.1	10.4
	$\bar{y}_{22}$	95.1	7.45	95.4	7.39
	$\bar{y}_{31}$	89.9***	9.84	94.1	10.5
	$\bar{y}_{32}$	94.1	7.50	95.6	7.42

*** significant at 0.1% level, ** significant at 1% level,

* significant at 5% level

Table 4.6 *mGREG estimates of* $\bar{y}_{1d}$ *in domains 1 and 2 when the same auxiliary variable (Aux var) is used in the numerator and the denominator.*

		Domain 1				*Domain 2*			
Case	*Aux var*	*Bias*	S^2 $\times 10^3$	$\bar{\hat{V}}$ $\times 10^3$	*CI95 %*	*Bias*	S^2 $\times 10^3$	$\bar{\hat{V}}$ $\times 10^3$	*CI95 %*
A)	x_{dk}	0.002	8.70	$7.88^{\times}$	88.6	-0.003	2.69	$3.05^{\times\times}$	95.1
	z_{dk}	0.003	8.56	$7.80^{\times}$	88.6	0.000	2.65	$3.00^{\times\times\times}$	95.8
	x_{dk}, z_{dk}	0.003	8.08	$8.91^{\times}$	95.6	-0.002^{*}	0.960	0.964	94.9
B)	x_{dk}	0.003	8.52	$7.69^{\times}$	88.3	-0.003	2.41	$2.83^{\times\times\times}$	95.9
	z_{dk}	0.004	8.53	$7.74^{\times}$	88.5	0.000	2.55	$2.96^{\times\times\times}$	95.9
	x_{dk}, z_{dk}	0.005	9.67	$11.64^{\times\times\times}$	95.1	-0.001	0.827	$0.916^{\times}$	96.4^{*}

Test of bias
*** significant at 0.1% level, ** significant at 1% level, * significant at 5% level

Test of $\bar{\hat{V}}(\hat{\bar{y}}_{id}) - S^2(\hat{\bar{y}}_{id})$
xxx significant at 0.1% level, xx significant at 1% level, x significant at 5% level

Fig 4.1-4.8 *Distribution of* 1,000 *estimated means,* $\hat{\bar{y}}_{1d}$ *and variances,* $\hat{V}(\hat{\bar{y}}_{1d})$, *d*=1, 2, *Case=A, when Mak & Li and the mGREG estimators are used.*

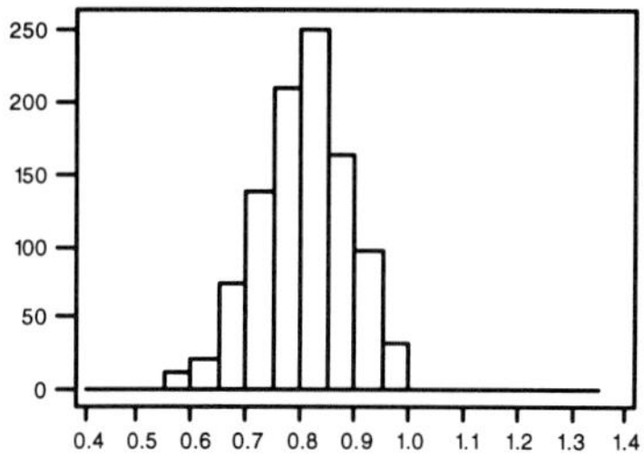

Fig 4.1 *Mak & Li,* $\hat{\bar{y}}_{11}$

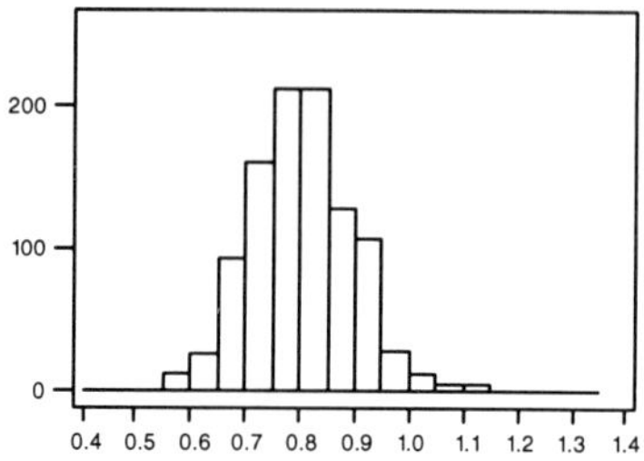

Fig 4.2 *mGREG,* $\hat{\bar{y}}_{11}$

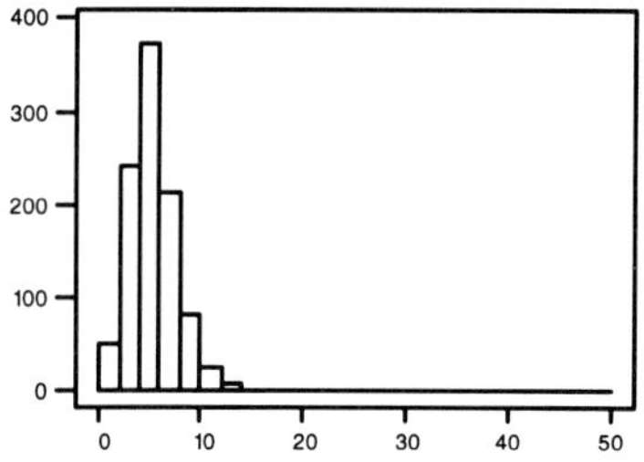

Fig 4.3 *Mak & Li,* $\hat{V}(\hat{\bar{y}}_{11})\times 10^3$

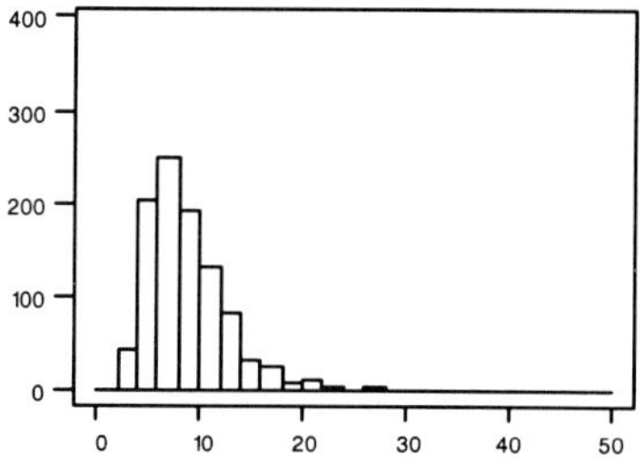

Fig 4.4 *mGREG,* $\hat{V}(\hat{\bar{y}}_{11})\times 10^3$

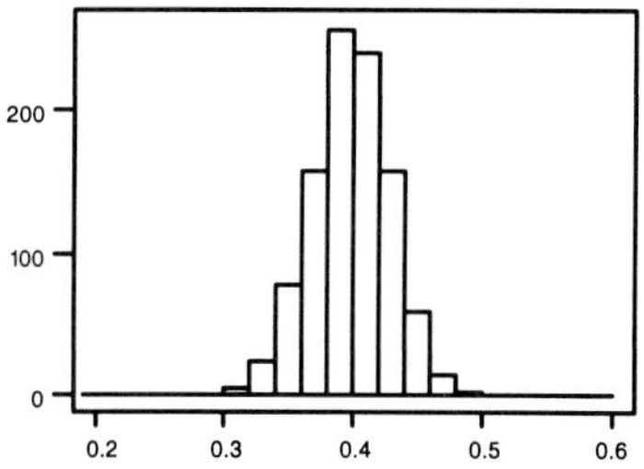

Fig 4.5 *Mak & Li,* $\hat{\bar{y}}_{12}$

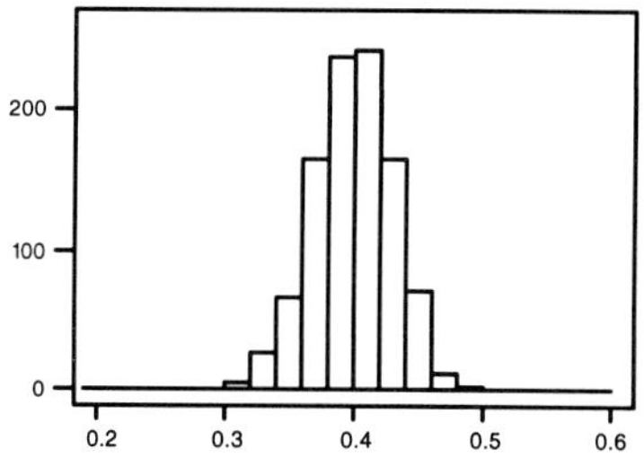

Fig 4.6 *mGREG,* $\hat{\bar{y}}_{12}$

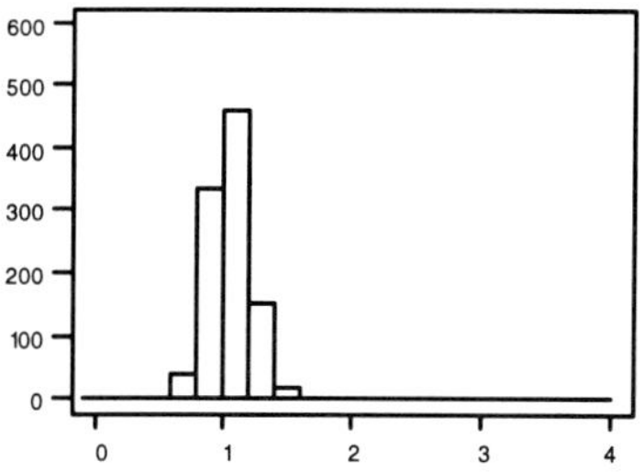

Fig 4.7 *Mak & Li,* $\hat{V}(\hat{\bar{y}}_{12})\times 10^3$

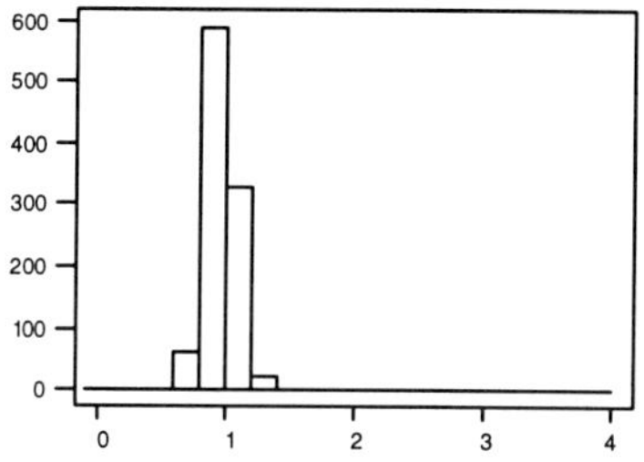

Fig 4.8 *mGREG,* $\hat{V}(\hat{\bar{y}}_{12})\times 10^3$

Fig 4.9-4.16 *Distribution of* 1,000 *estimated means,* $\hat{\bar{y}}_{2d}$ *and variances,* $\hat{V}(\hat{\bar{y}}_{2d})$, *d*=1, 2, *Case=A, when Mak & Li and the mGREG estimators are used.*

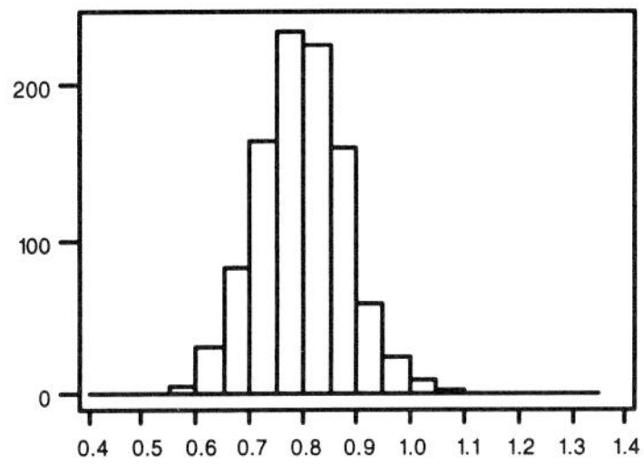

Fig 4.9 *Mak & Li,* $\hat{\bar{y}}_{21}$

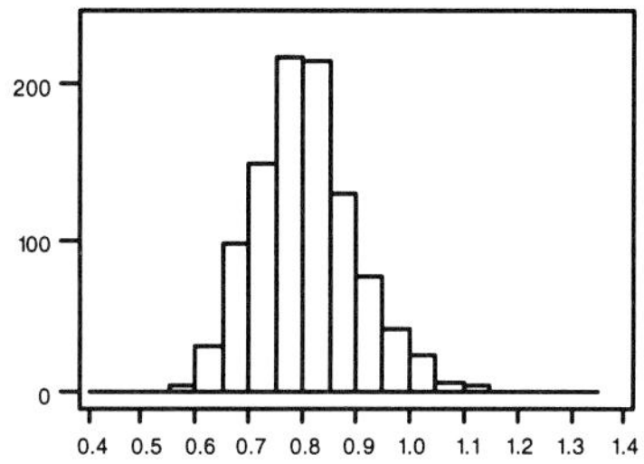

Fig 4.10 *mGREG,* $\hat{\bar{y}}_{21}$

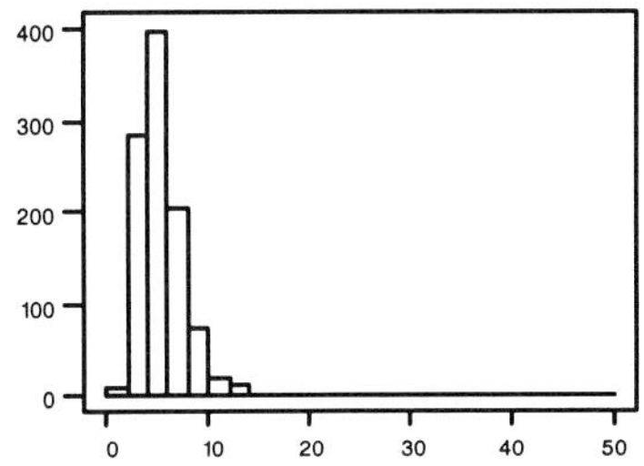

Fig 4.11 *Mak & Li,* $\hat{V}(\hat{\bar{y}}_{21})\times 10^3$

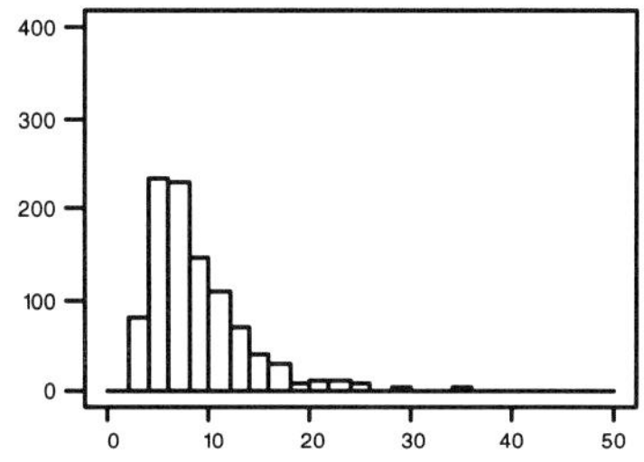

Fig 4.12 *mGREG,* $\hat{V}(\hat{\bar{y}}_{21})\times 10^3$

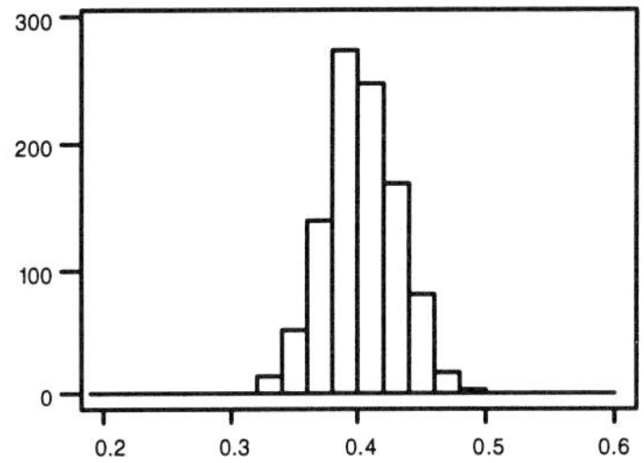

Fig 4.13 *Mak & Li,* $\hat{\bar{y}}_{22}$

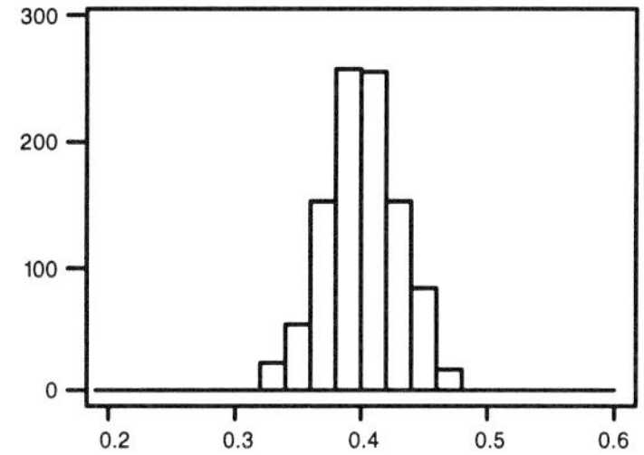

Fig 4.14 *mGREG,* $\hat{\bar{y}}_{22}$

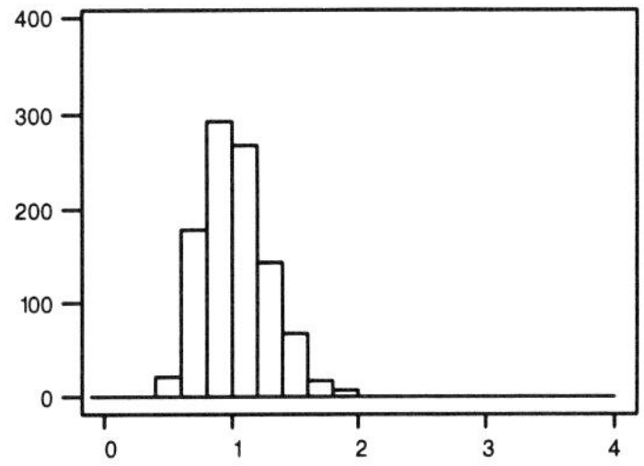

Fig 4.15 *Mak & Li,* $\hat{V}(\hat{\bar{y}}_{22})\times 10^3$

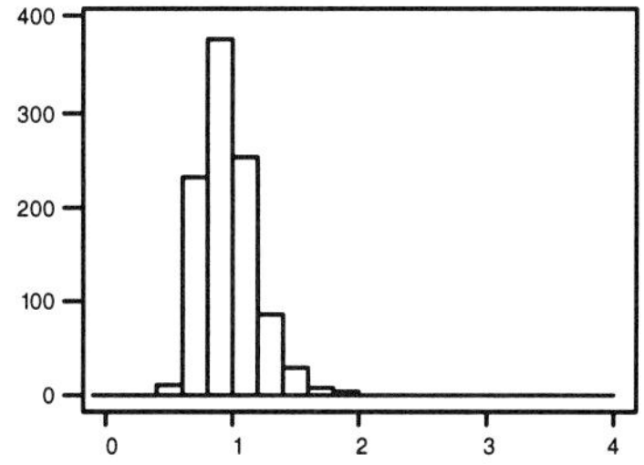

Fig 4.16 *mGREG,* $\hat{V}(\hat{\bar{y}}_{22})\times 10^3$

Fig 4.17-4.24 *Distribution of 1,000 estimated means,* $\hat{\bar{y}}_{3d}$ *and variances,* $\hat{V}(\hat{\bar{y}}_{3d})$*,* d=1, 2, *Case=A, when Mak & Li and the mGREG estimators are used.*

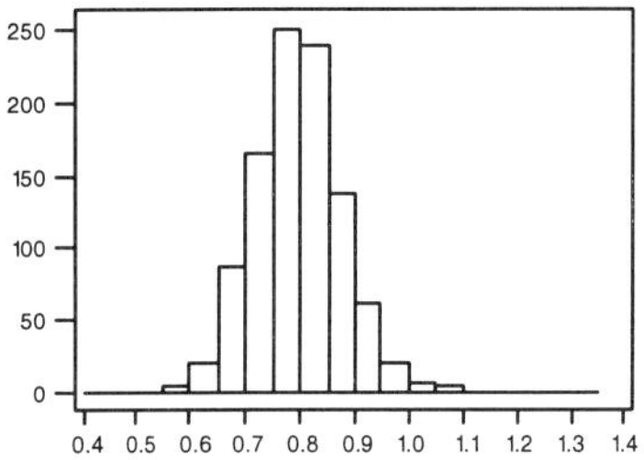

Fig 4.17 *Mak & Li,* $\hat{\bar{y}}_{31}$

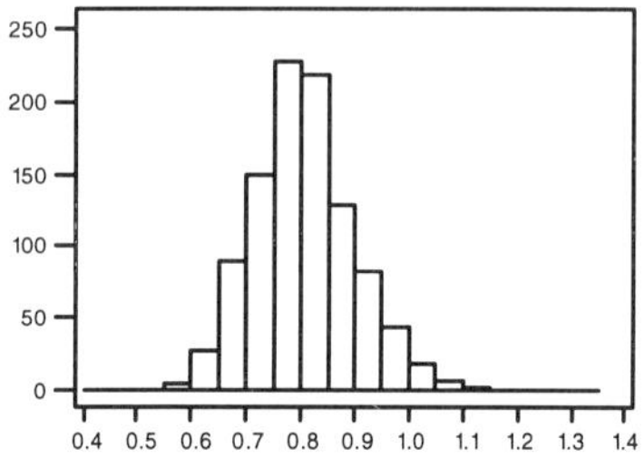

Fig 4.18 *mGREG,* $\hat{\bar{y}}_{31}$

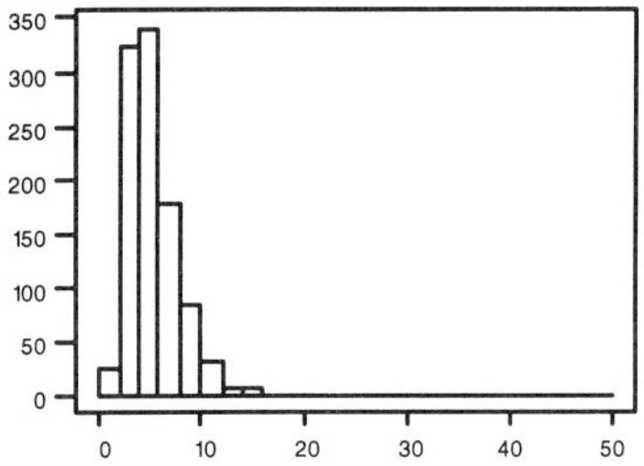

Fig 4.19 *Mak & Li,* $\hat{V}(\hat{\bar{y}}_{31})\times 10^3$

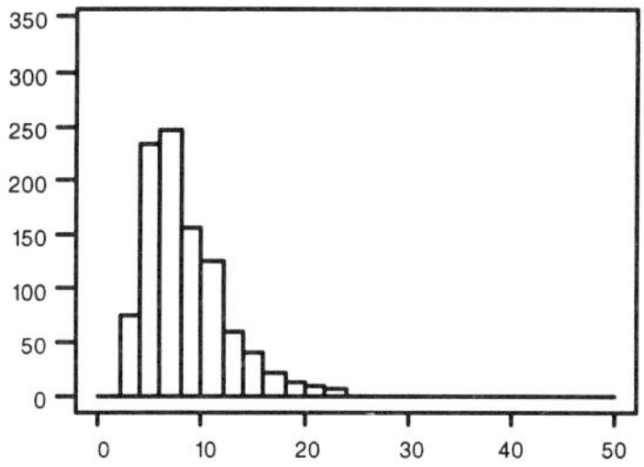

Fig 4.20 *mGREG,* $\hat{V}(\hat{\bar{y}}_{31})\times 10^3$

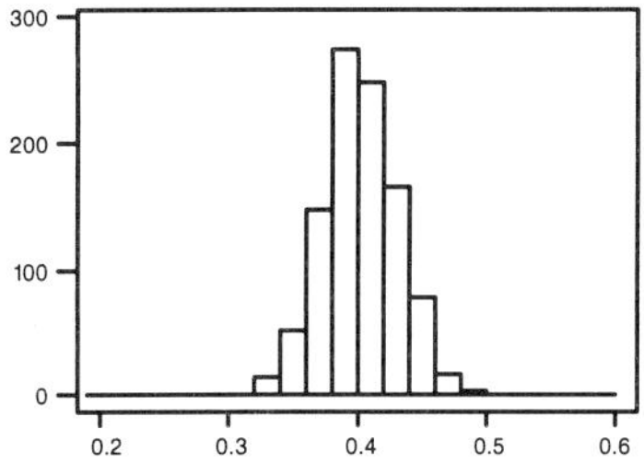

Fig 4.21 *Mak & Li,* $\hat{\bar{y}}_{32}$

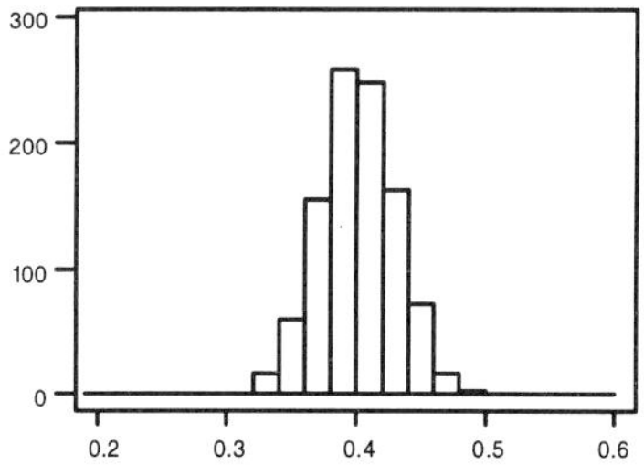

Fig 4.22 *mGREG,* $\hat{\bar{y}}_{32}$

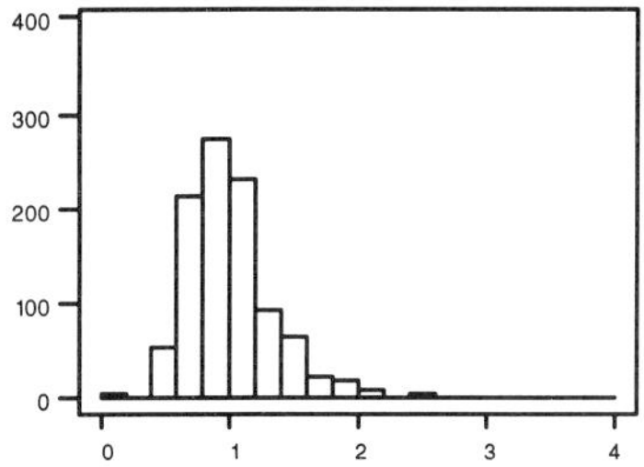

Fig 4.23 *Mak & Li,* $\hat{V}(\hat{\bar{y}}_{32})\times 10^3$

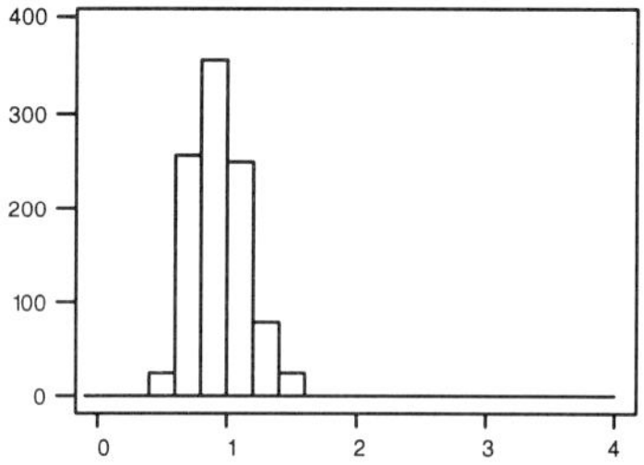

Fig 4.24 *mGREG,* $\hat{V}(\hat{\bar{y}}_{32})\times 10^3$

Fig 4.25-4.32 *Distribution of* 1,000 *estimated means,* $\hat{\hat{\bar{y}}}_{1d}$ *and variances,* $\hat{V}(\hat{\hat{\bar{y}}}_{1d})$, *d*=1, 2, *Case=B, when Mak & Li and the mGREG estimators are used.*

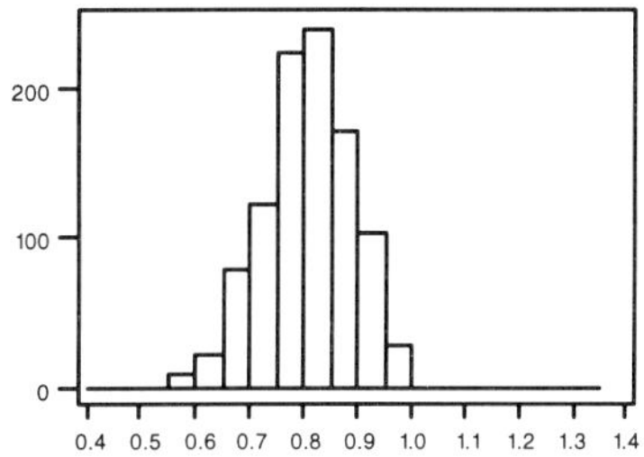

Fig 4.25 *Mak & Li,* $\hat{\hat{\bar{y}}}_{11}$

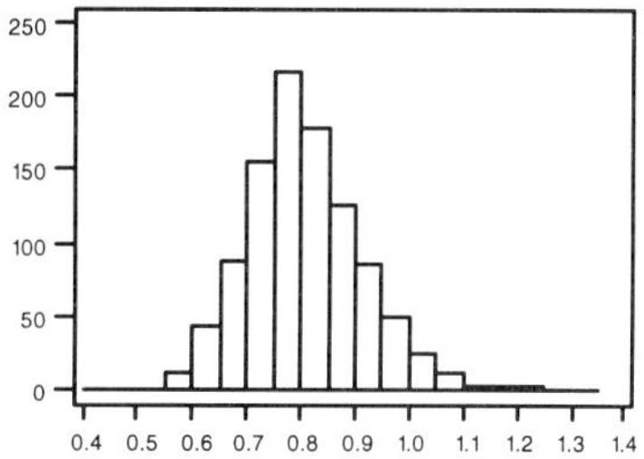

Fig 4.26 *mGREG,* $\hat{\hat{\bar{y}}}_{11}$

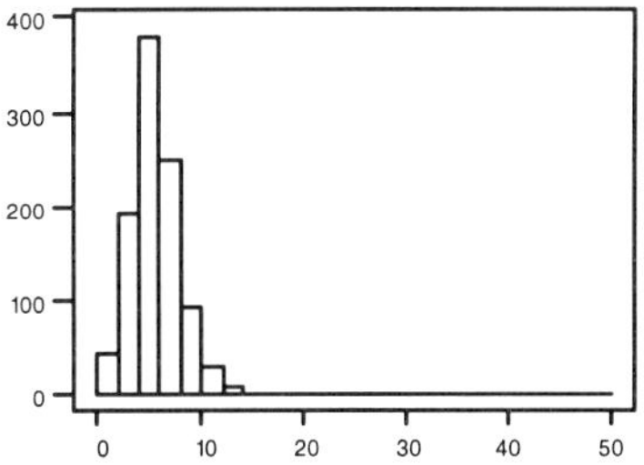

Fig 4.27 *Mak & Li,* $\hat{V}(\hat{\hat{\bar{y}}}_{11})\times 10^3$

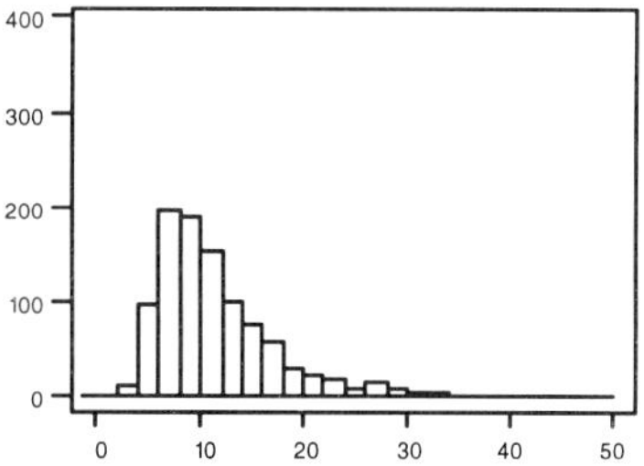

Fig 4.28 *mGREG,* $\hat{V}(\hat{\hat{\bar{y}}}_{11})\times 10^3$

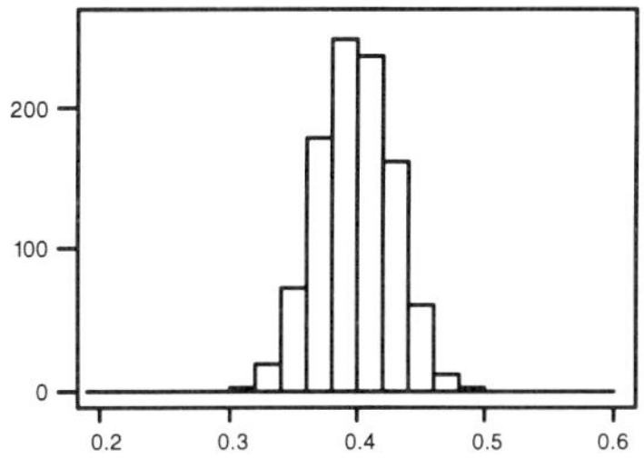

Fig 4.29 *Mak & Li,* $\hat{\hat{\bar{y}}}_{12}$

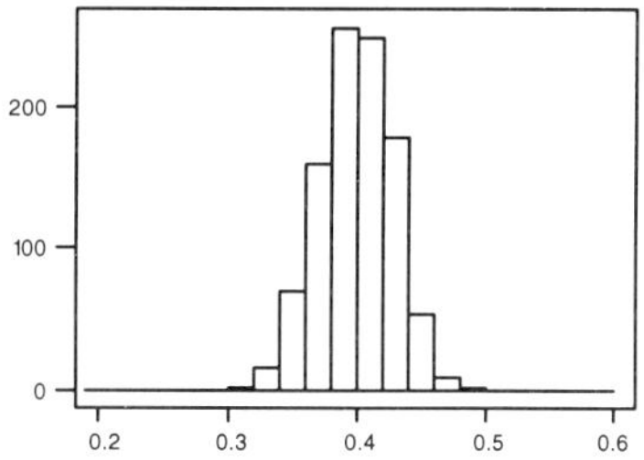

Fig 4.30 *mGREG,* $\hat{\hat{\bar{y}}}_{12}$

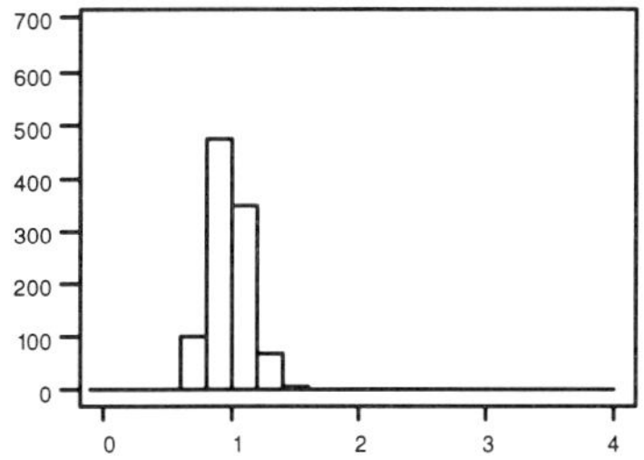

Fig 4.31 *Mak & Li,* $\hat{V}(\hat{\hat{\bar{y}}}_{12})\times 10^3$

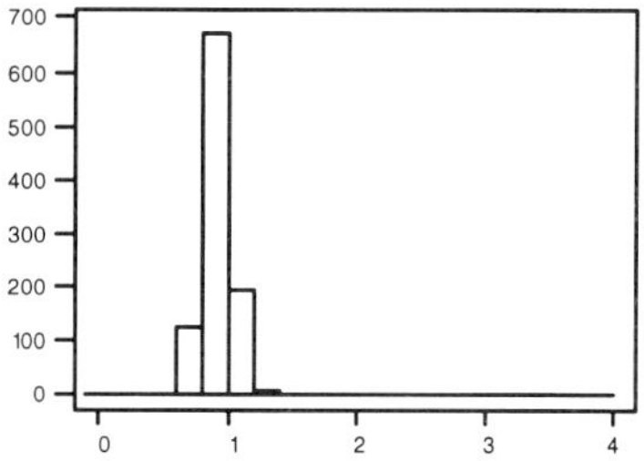

Fig 4.32 *mGREG,* $\hat{V}(\hat{\hat{\bar{y}}}_{12})\times 10^3$

Fig 4.33-4.40 *Distribution of* 1,000 *estimated means,* $\hat{\bar{y}}_{2d}$ *and variances,* $\hat{V}(\hat{\bar{y}}_{2d})$, *d*=1, 2, *Case=B, when Mak & Li and the mGREG estimators are used.*

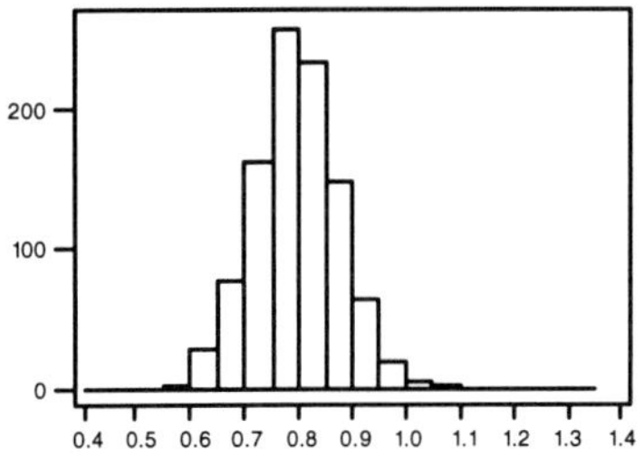

Fig 4.33 *Mak & Li,* $\hat{\bar{y}}_{21}$

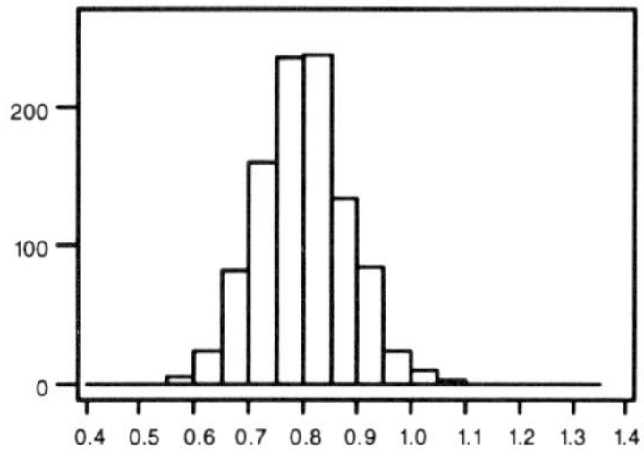

Fig 4.34 *mGREG,* $\hat{\bar{y}}_{21}$

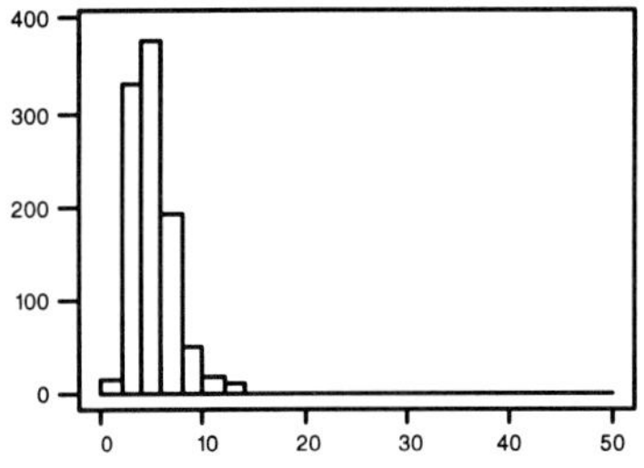

Fig 4.35 *Mak & Li,* $\hat{V}(\hat{\bar{y}}_{21})\times 10^3$

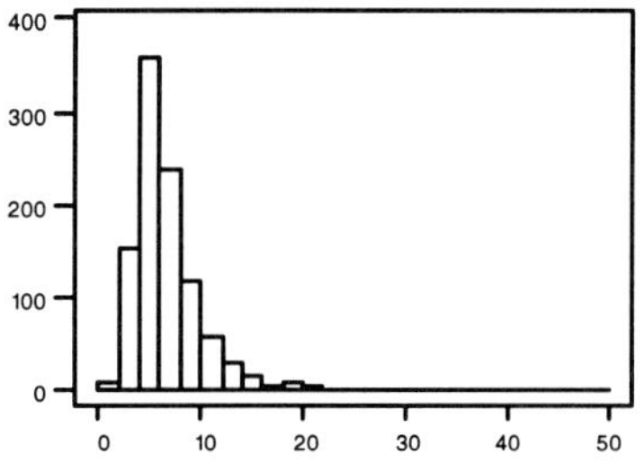

Fig 4.36 *mGREG,* $\hat{V}(\hat{\bar{y}}_{21})\times 10^3$

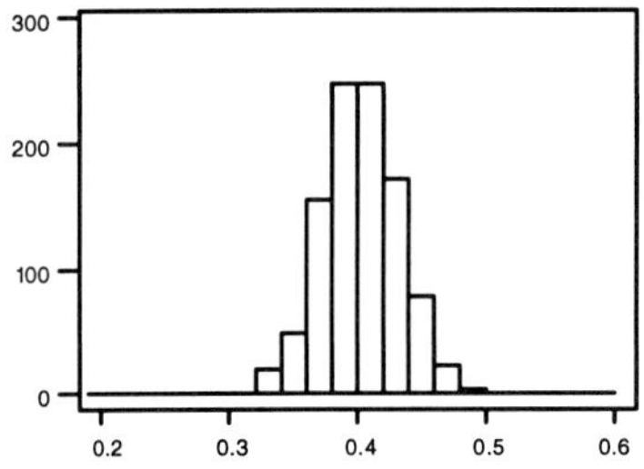

Fig 4.37 *Mak & Li,* $\hat{\bar{y}}_{22}$

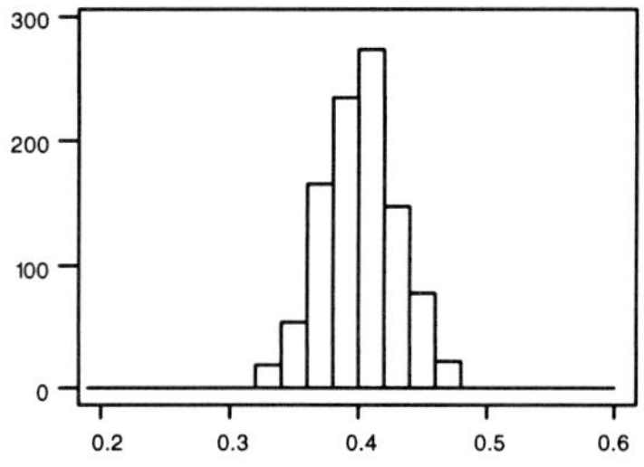

Fig 4.38 *mGREG,* $\hat{\bar{y}}_{22}$

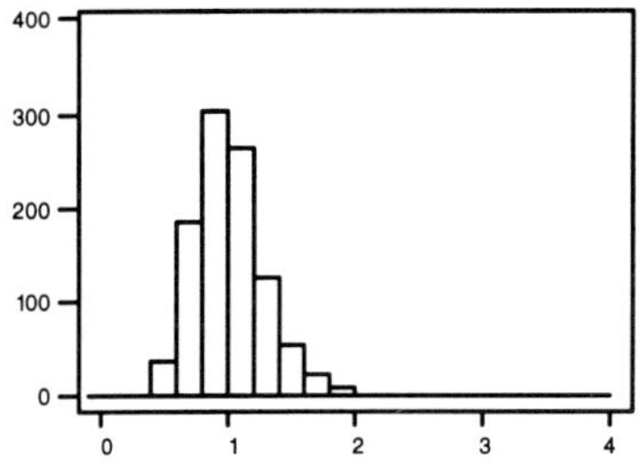

Fig 4.39 *Mak & Li,* $\hat{V}(\hat{\bar{y}}_{22})\times 10^3$

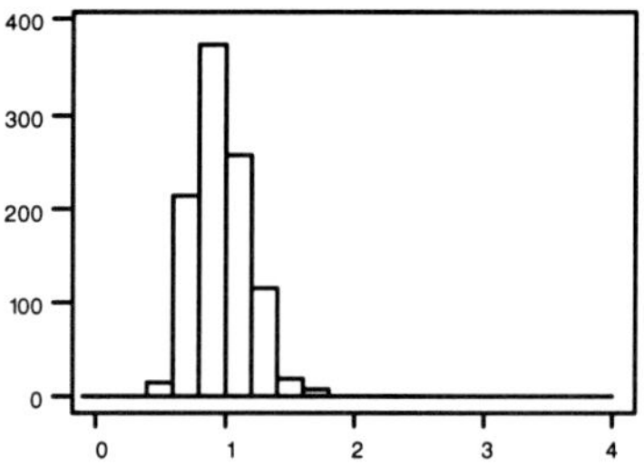

Fig 4.40 *mGREG,* $\hat{V}(\hat{\bar{y}}_{22})\times 10^3$

Fig 4.41-4.48 *Distribution of* 1,000 *estimated means,* $\hat{\bar{y}}_{3d}$ *and variances,* $\hat{V}(\hat{\bar{y}}_{3d})$, d=1, 2, *Case=B, when Mak & Li and the mGREG estimators are used.*

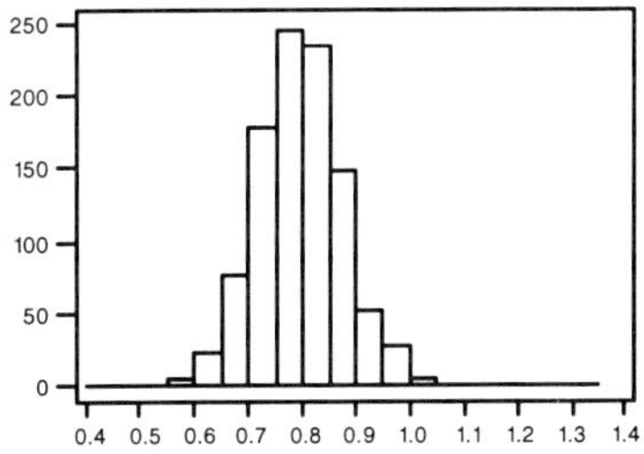

Fig 4.41 *Mak & Li,* $\hat{\bar{y}}_{31}$

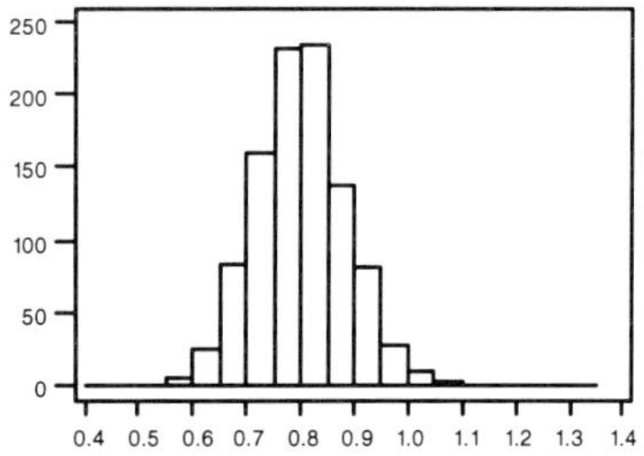

Fig 4.42 *mGREG,* $\hat{\bar{y}}_{31}$

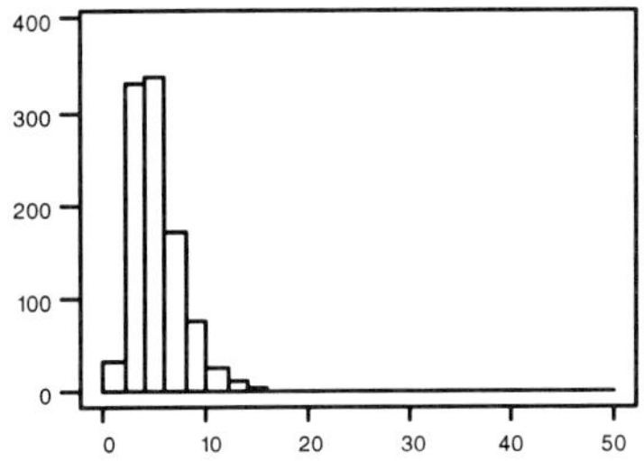

Fig 4.43 *Mak & Li,* $\hat{V}(\hat{\bar{y}}_{31})\times 10^3$

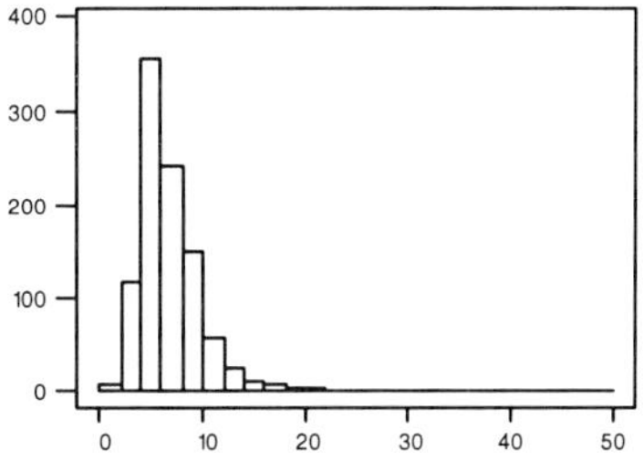

Fig 4.44 *mGREG,* $\hat{V}(\hat{\bar{y}}_{31})\times 10^3$

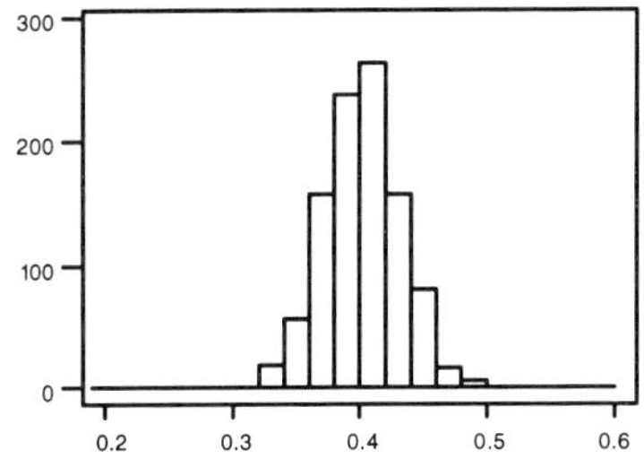

Fig 4.45 *Mak & Li,* $\hat{\bar{y}}_{32}$

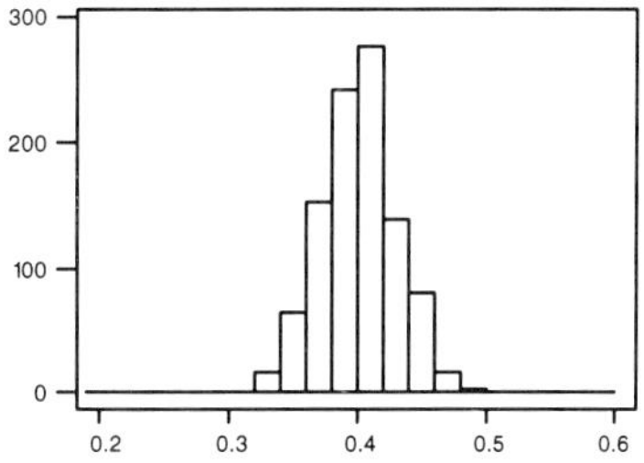

Fig 4.46 *mGREG,* $\hat{\bar{y}}_{32}$

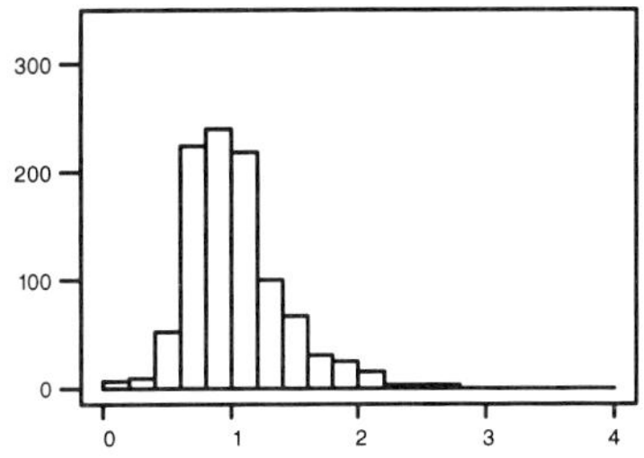

Fig 4.47 *Mak & Li,* $\hat{V}(\hat{\bar{y}}_{32})\times 10^3$

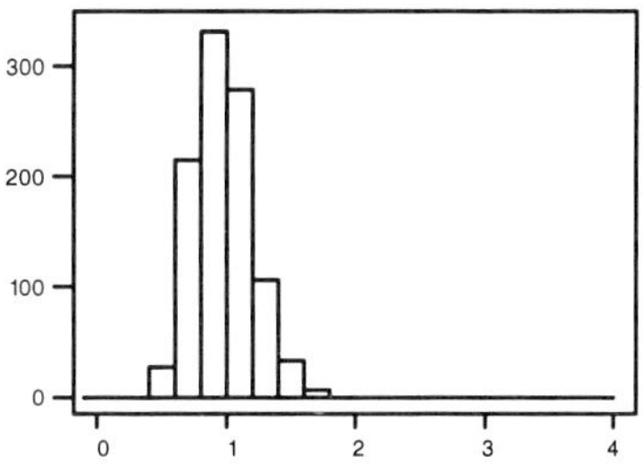

Fig 4.48 *mGREG,* $\hat{V}(\hat{\bar{y}}_{32})\times 10^3$

Fig 4.49-4.56 *Distribution of* 1,000 *estimated means,* $\hat{\bar{y}}_{1d}$ *and variances,* $\hat{V}(\hat{\bar{y}}_{1d})$, *d*=1, 2, *Case=A, mGREG and the same auxiliary variable* (*Aux var*) *in the numerator and the denominator.*

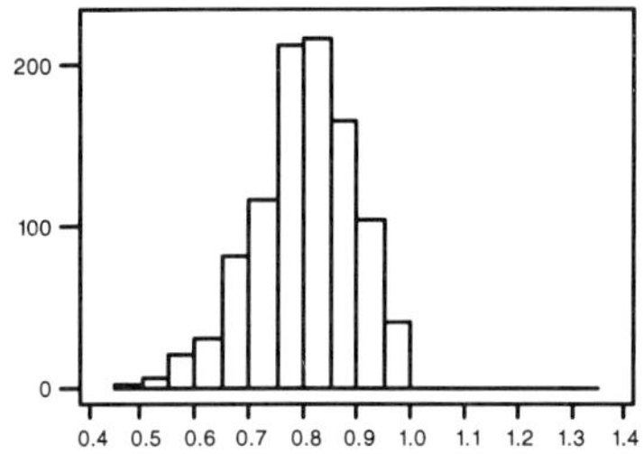

Fig 4.49 $\hat{\bar{y}}_{11}$, *Aux var=x*

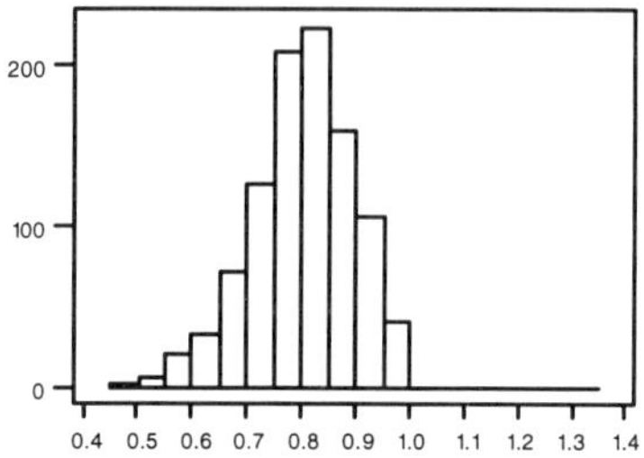

Fig 4.50 $\hat{\bar{y}}_{11}$, *Aux var=z*

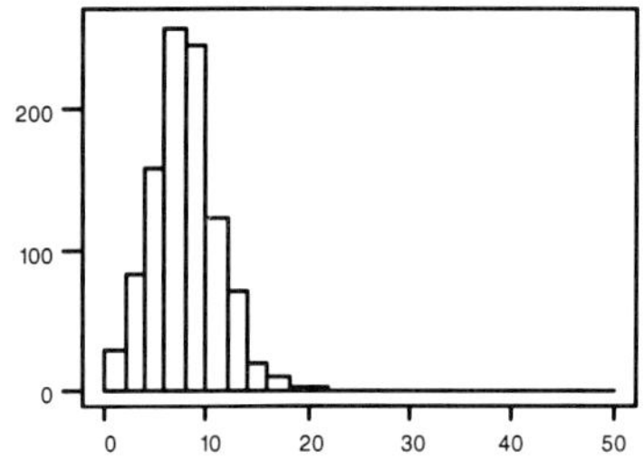

Fig 4.51 $\hat{V}(\hat{\bar{y}}_{11})\times 10^3$, *Aux var=x*

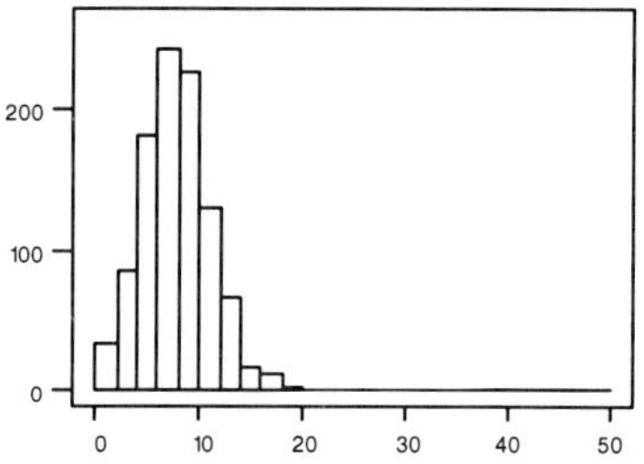

Fig 4.52 $\hat{V}(\hat{\bar{y}}_{11})\times 10^3$, *Aux var=z*

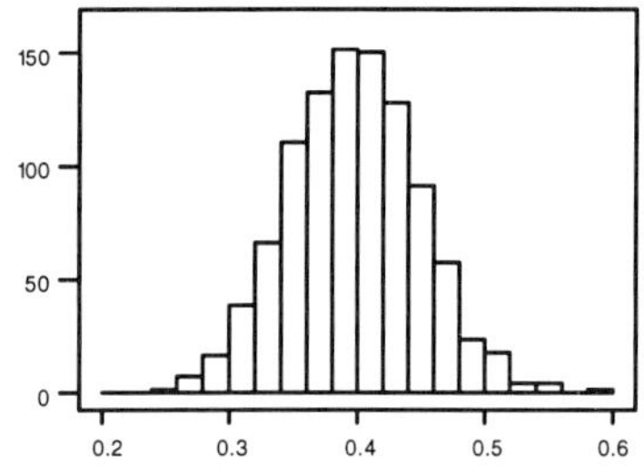

Fig 4.53 $\hat{\bar{y}}_{12}$, *Aux var=x*

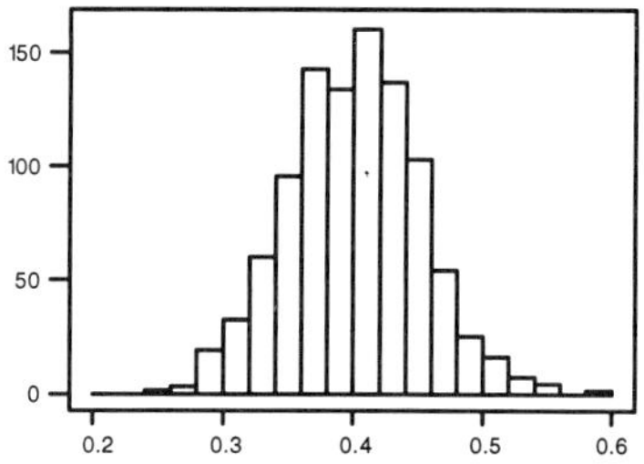

Fig 4.54 $\hat{\bar{y}}_{12}$, *Aux var=z*

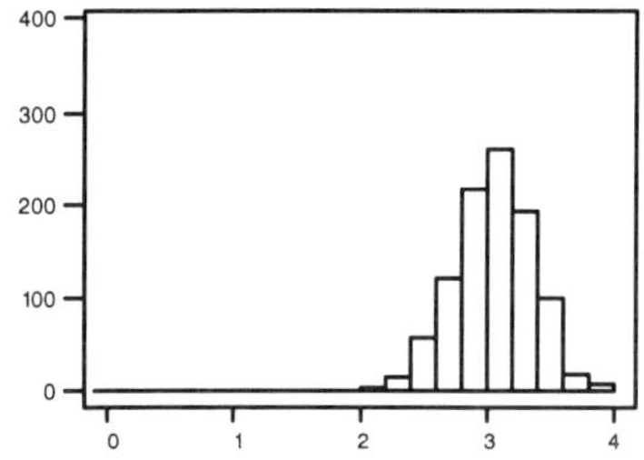

Fig 4.55 $\hat{V}(\hat{\bar{y}}_{12})\times 10^3$, *Aux var=x*

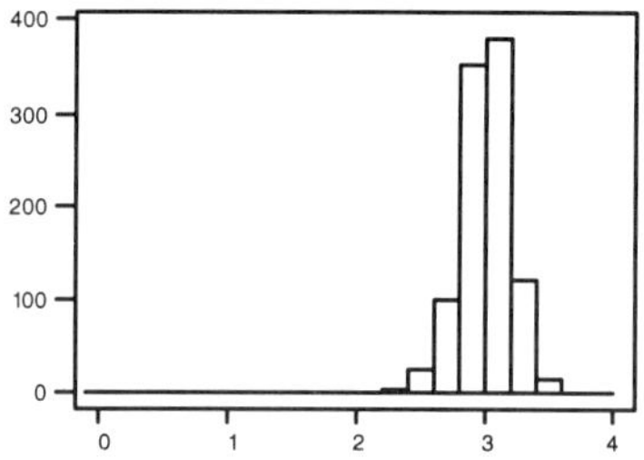

Fig 4.56 $\hat{V}(\hat{\bar{y}}_{12})\times 10^3$, *Aux var=z*

Fig 4.57-4.64 *Distribution of* 1,000 *estimated means,* $\hat{\bar{y}}_{1d}$ *and variances,* $\hat{V}(\hat{\bar{y}}_{1d})$, *d*=1, 2, *Case=B, mGREG and the same auxiliary variable* (*Aux var*) *in the numerator and the denominator.*

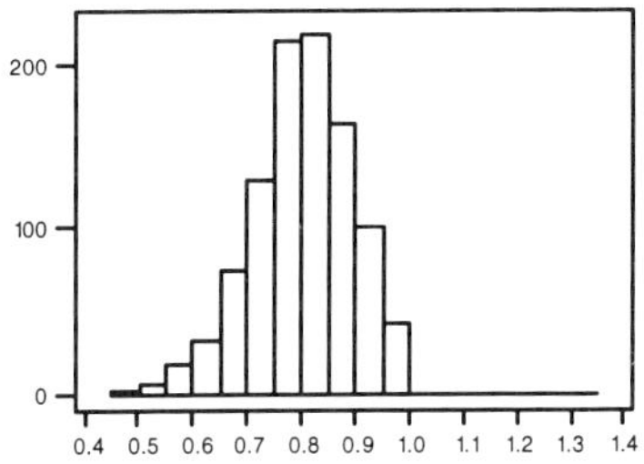

Fig 4.57 $\hat{\bar{y}}_{11}$, *Aux var=x*

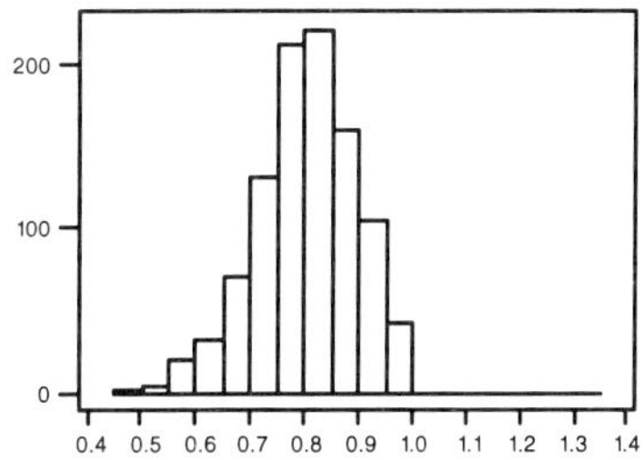

Fig 4.58 $\hat{\bar{y}}_{11}$, *Aux var=z*

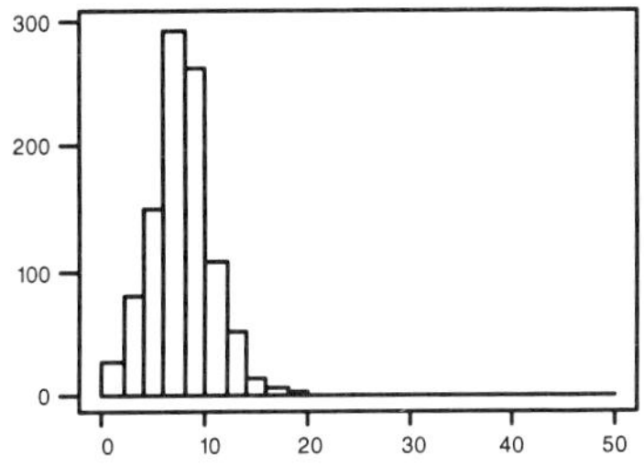

Fig 4.59 $\hat{V}(\hat{\bar{y}}_{11})\times 10^3$, *Aux var=x*

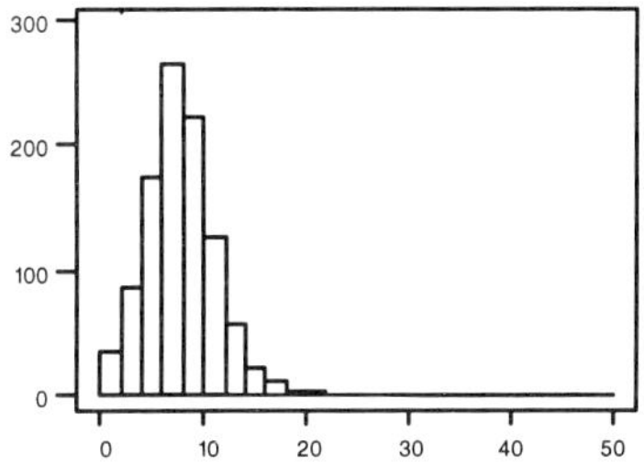

Fig 4.60 $\hat{V}(\hat{\bar{y}}_{11})\times 10^3$, *Aux var=z*

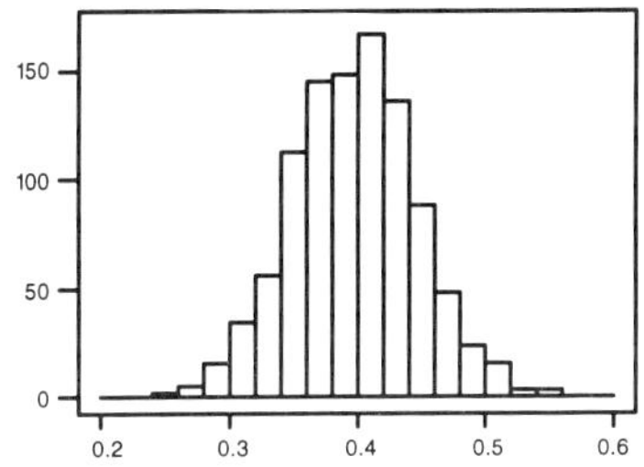

Fig 4.61 $\hat{\bar{y}}_{12}$, *Aux var=x*

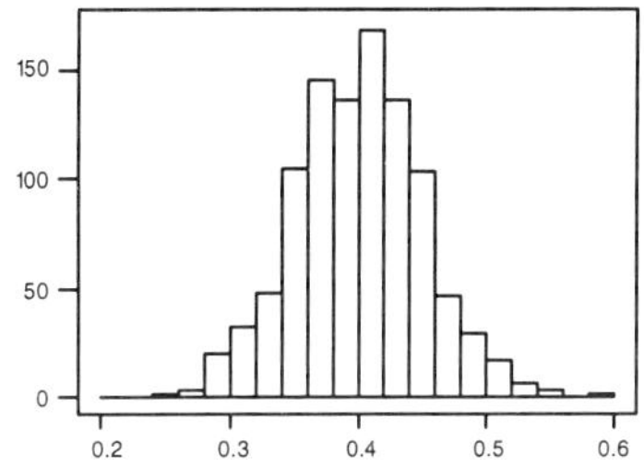

Fig 4.62 $\hat{\bar{y}}_{12}$, *Aux var=z*

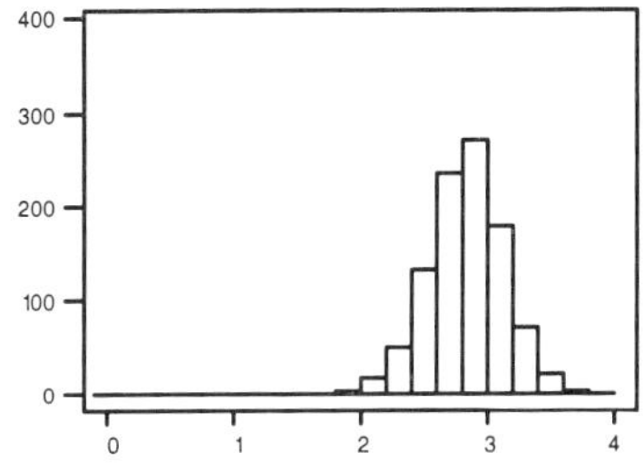

Fig 4.63 $\hat{V}(\hat{\bar{y}}_{12})\times 10^3$, *Aux var=x*

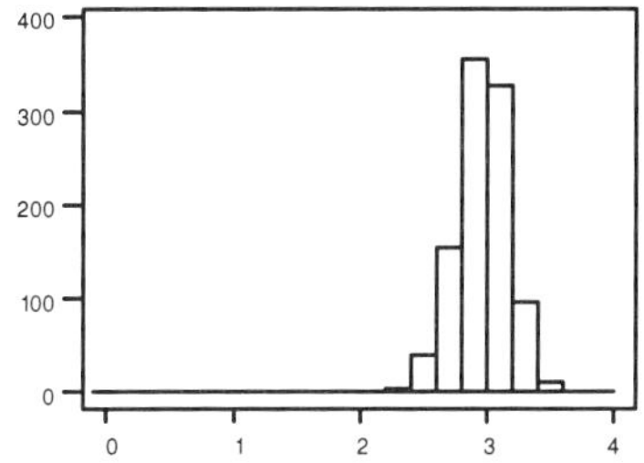

Fig 4.64 $\hat{V}(\hat{\bar{y}}_{12})\times 10^3$, *Aux var=z*

CHAPTER 5

Misclassification and Measurement Errors when *SI* Designs are Used

When there are classification errors as well as measurement errors in a survey, none of Selén's, Mak's and Mak & Li's estimators can be used, even in the simple situation when an *SI* design is used in both phases. Selén (1986) suggested a technique to handle the situation with both types of errors, but he did not give an explicit expression for the estimator.

The estimator proposed by Mak & Li (1988) is only applicable when (*i*) $p_a(\cdot)$ is a *SIR* design (or an *SI* design with a small sample fraction) and $p(\cdot \mid s_a)$ is an *SI* design, (*ii*) the parameters of interest are subgroup means and (*iii*) there are no errors in the y-values when the procedure F is used.

Although the estimator presented by Mak & Li cannot be used when $y_k \neq x_k$, i.e. when there are measurement errors in the observed x_k-values, it is possible to derive an estimator that allows measurement errors as well as classification errors, along the same lines as Mak & Li does. By casting the estimation formulas into the same form as (4.12)-(4.15) and defining $x_{dk}=z_{dk}x_k$ and $y_{dk}=\zeta_{dk}y_k$, the same formal estimator can be used as when $y_k=x_k$ for every k.

Andersson (1992) suggested that the *m*GREG estimator should be used when there are both classification and measurement errors in the data since that situation can be handled in a straightforward way and under any practical design.

5.1 The Simultaneous Effects of Misclassification and Measurement Errors — A Simulation Study

A simulation study was carried out to investigate the sampling properties of the *m*GREG estimator and the modified Mak & Li estimator with respect to bias, variance and distributional properties. The simulation study used the same artificial data as in chapter 4. The programming was done by the use of APL*PLUS 7.0, STSC Inc. and an IBM PS/2 model 70 computer.

5.1.1 The Simulation Study

The same population as in chapter 4 and the same misclassification probabilities as in Cases A) and B) were used in the simulation.

The sample size in the first phase was n_a=500, and the sample size in the second phase was n=100. Both samples were drawn without replacement.

Each unit in s_a was classified into class 1 or class 2 by using two different sets of classification probabilities, Cases A) and B). The measurement process was simulated in the following way. Variable y_1, which is a 0-1 variable, was assigned its true value with probability 0.9, i.e. $\Pr(x_k=1 \mid y_{1k}=1)=\Pr(x_k=0 \mid y_{1k}=0)=0.9$, $k \in s_a$ and thus

$$\mu_{xk}^{(1)} = \mu_{xk} = \begin{cases} 0.9 & y_{1k} = 1 \\ 0.1 & y_{1k} = 0 \end{cases} \quad (5.1)$$

and $\sigma_{xk}^2 = 0.09$.

The measurement errors in y_2 and y_3 were generated by adding a value from a normal distributed variable δ with expectation zero, i.e. $x_k=y_k+\delta_k$, $k \in s_a$, and the expectation of x_k is $\mu_{xk}^{(1)}=y_k$. The standard deviation of δ was chosen in two ways: 1) $\sigma_\delta=$ $=0.1y_k$ and 2) $\sigma_\delta=0.05$.

In each Cases A) and B), 1,000 samples were taken. For each sample j (j=1, ..., 1,000) the estimates $\hat{\bar{y}}_{idj}$ (i=1, 2, 3, d=1, 2) for variable i in domain d were calculated for each proposed estimator. Furthermore, the estimated variances $\hat{V}(\hat{\bar{y}}_{idj})$ and the intervals $\hat{\bar{y}}_{idj} \pm 1.96\sqrt{\hat{V}(\hat{\bar{y}}_{idj})}$ were calculated, and it was checked whether the interval covered the true mean $\bar{y}_{id}$ or not.

5.1.2 Results of the Simulation Study

The distributions of 1,000 estimates $\hat{\bar{y}}_{id}$ and $\hat{V}(\hat{\bar{y}}_{id})$ i=1, 2, 3, d=1, 2, Cases A) and B), for Mak & Li's and the mGREG estimators are shown in figures 5.1-5.48. Only measurement 1) where $\sigma_\delta=0.1y_k$ is shown because the distributions were similar to those in measurement 2) where $\sigma_\delta=0.05$.

Bias

The hypothesis that $E\left[\hat{\bar{y}}_{id}\right]-\bar{y}_{id}=0$ for each i and d is tested by (4.19) in the same way as in section 4.3.3. The results of the tests are given in table 5.1.

NOTE: Both the modified Mak & Li's estimator and *m*GREG gave inadmissible estimates of $\bar{y}_{11}$, i.e. estimates that fell outside (0, 1), in some samples. When an inadmissible estimate was obtained it was set at 1; no negative estimates could be obtained in this case. The proportions of inadmissible estimates in Case A) were 1.6% for the modified Mak & Li's estimator and 5.9% for *m*GREG, in Case B) the proportions were 2.1% and 7.2% respectively. □

The observed biases are significant at the 0.1% level when $\bar{y}_{11}$ is estimated by Mak & Li's estimator in Cases A) and B), while the observed bias of the *m*GREG estimator is significant at the 1% level in these cases. The biases are probably due to the inadmissible estimates. In all other cases the observed biases are not significant or significant only at the 5% level. The largest observed values of *ABR* are 0.14 and 0.12. Both values were obtained for the estimate of $\bar{y}_{11}$ when Mak & Li's estimator was used in Cases A) and B); in all other cases the values of *ABR* were between 0.00 and 0.09. Disregarding the failure of Mak & Li's estimator when inadmissible estimates occur, both proposed estimators are considered useful as far as bias is concerned.

Comparison of Variance Estimators and Coverage Rates

The variances of the estimated values calculated by Mak & Li's and the *m*GREG estimators are compared in the same way as in section 4.3.3. The upper and lower limits for the variance of the estimated mean values $S^2(\hat{\bar{y}}_{id})$ were calculated according to (4.22) and the confidence intervals for $\bar{\hat{V}}(\hat{\bar{y}}_{id})$ were calculated according to (4.23). The results are given in table 5.2 for Mak & Li's estimator and in table 5.3 for the *m*GREG estimator.

The differences $\bar{\hat{V}}(\hat{\bar{y}}_{id}) - S^2(\hat{\bar{y}}_{id})$ are jackknifed in the same way as in chapter 4. When Mak & Li's estimator is used, the jackknifed differences are significant at the 0.1% or 1% levels for each variable and domain in Case A). In Case B) the differences are significant at the 0.1% or 1% level except for variables y_2 and y_3 in domain 2. This result is an indicator of a bias in the modified Mak & Li's variance estimator.

When the *m*GREG estimator is used, only the differences between the variances for variable y_1 in domain 1 are significant (0.1% level), probably due to the inadmissible estimates, which indicates that the bias in *m*GREG's variance estimator may be considered negligible in most of the studied cases.

The observed variances $S^2(\hat{\bar{y}}_{id})$ are usually smaller when Mak & Li's estimator is used than when the *m*GREG estimator is used. The jackknife technique is used to test whether the $S^2(\hat{\bar{y}}_{id})$ differs significantly between Mak & Li's and the *m*GREG estimators or not. In Case A) the jackknifed differences are significant at the 0.1% level in domain 1 for all three variables in Case A), but in domain 2 the differences are only significant at the 5% level for variable y_2. In Case B) the differences are significant at the 0.1% level in domain 1 for variable y_1 and at the 5% level for y_2 and the 1% level for y_3. In domain 2 the differences are significant at the 5% level for variables y_2 and y_3.

The observed differences between the *cves* calculated for Mak & Li's and the *m*GREG estimators are small. The observed coverage rates of the estimated confidence intervals are too small in domain 1 when Mak & Li's estimator is used, all coverage rates differing significantly from 95% at the 0.1% level. In domain 2, three estimated intervals are significant. When the *m*GREG estimator is used, only one of the estimated intervals differs significantly from 95%. The *cves* and the coverage rates are given in table 5.4.

The Problem of Inadmissible Values

The estimates of $\bar{y}_{11}$ larger than 1 were set at 1 in the same way as in chapter 4. When Mak & Li's estimator is used, there does not seem to be any remedy for the inadmissible values other than to replace these estimates with admissible ones. However, an *m*GREG-based estimator that does not give inadmissible estimates of the mean value can be constructed in the same way as in section 4.3.3 by using the same auxiliary variable in the denominator and the numerator.

A simulation study was carried out to study the sampling properties of the two choices of auxiliary variables when $\bar{y}_{11}$and $\bar{y}_{12}$ were estimated by the *m*GREG estimator in Cases A) and B). The results of that study are shown in table 5.5 and figures 5.49-5.64.

When the same auxiliary variable, x or z, was used in the denominator and the numerator, no inadmissible estimates were observed. The results are summarised under the following points:

i) The observed biases are not significant in any of the studied cases.

ii) The differences $\overline{\hat{V}}(\hat{\bar{y}}_{id})-S^2(\hat{\bar{y}}_{id})$ are not significant in domains 1 and 2.

iii) Due to the decreased values of the estimated variances $\hat{V}$ in domain 1 the observed coverage rates of the estimated confidence intervals are far below the intended 95%, while they are still non-significant in domain 2.

iv) It does not seem to matter if x or z is used as auxiliary variables in the studied cases, the variances and the coverage rates are about the same.

Instead of *always* using the same auxiliary variable in the denominator and the numerator when *m*GREG is used, the following strategy may be applied: (*i*) use different auxiliary variables in the denominator and numerator as long as the estimates are within (0, 1), (*ii*) when an inadmissible estimate is encountered, use the same auxiliary variable in the denominator and the numerator.

A simulation study was carried out to investigate the sampling properties of this strategy. Since there were no inadmissible estimates in domain 2, the results given in table 5.6 are only applicable to domain 1. The findings are summarised in the following points:

i) The observed biases are negligible in all cases.

ii) The differences $\overline{\hat{V}}(\hat{\bar{y}}_{id}) - S^2(\hat{\bar{y}}_{id})$ are significant at the 0.1% level in domains 1 and 2.

iii) None of the observed confidence intervals has a coverage rate that differs significantly from 95%

iv) The same result is obtained irrespective of which one of x and z is used as auxiliary variable in the estimation.

5.1.3 Conclusions from the Simulation Study

On the basis of the results of this simulation study the following conclusions can be drawn:

i) Bias

Mak & Li's estimator and the *m*GREG estimator seem to give estimates with negligible bias, except when inadmissible estimates of $\bar{y}_{11}$ are obtained.

ii) Variance

The observed variances of Mak & Li's estimates are in general smaller than those variances of the *m*GREG estimates, but the variance estimator proposed by Mak & Li seems to systematically underestimate the variances in domain 1. The bias in the variance estimator proposed for *m*GREG is negligible in the studied cases, at least when different auxiliary variables are used in the numerator and the denominator.

In domain 1 and 2 the distributions of the *m*GREG-estimates of the variances are more spread out compared to those of Mak & Li when variable y_1 is used. When variables y_2 and y_3 are used the distributions of the *m*GREG-estimates of the variances are more spread out compared to those of Mak & Li in domain 1. The opposite is at hand in domain 2 as can be seen from figures 5.1-5.48.

iii) Distributional properties

The coefficients of variation, *cve*, are about the same when Mak & Li's and the *m*GREG estimators are used. The coverage rates of the intended 95% confidence intervals differ due to the underestimates of the variances when Mak & Li's estimator is used; these rates are much lower than 95%, especially in domain 1. The intervals calculated for the *m*GREG estimator have approximately the intended coverage rates.

The conclusions that can be drawn from *i)-iii)* are that Mak & Li's approach is inferior due to a biased variance estimator that gives improper estimates of confidence intervals in many cases, while the *m*GREG estimator mainly behaves well regarding the criteria used in this study. The strategy of using different auxiliary variables in the denominator and the numerator for *m*GREG except when an inadmissible estimate is obtained seems to be reasonable.

Different first- and second-phase sample sizes and different measurement error characteristics have also been tried. The results are similar to those presented here.

Table 5.1 *Observed bias in the estimated mean for each of the proposed estimators*

		Case			
		A)		B)	
Measurement	*Parameter*	*Mak & Li*	*mGREG*	*Mak & Li*	*mGREG*
Pr(true)=0.9	$\bar{y}_{11}$	0.011^{***}	0.008^{**}	0.013^{***}	0.010^{**}
	$\bar{y}_{12}$	0.001	0.001	0.001	0.001
$\delta_k \sim N(0, 0.1y_k)$	$\bar{y}_{21}$	-0.001	0.007^{*}	0.001	0.004
	$\bar{y}_{22}$	0.002^{*}	0.000	0.002^{*}	0.000
	$\bar{y}_{31}$	-0.002	0.006^{*}	-0.001	0.003
	$\bar{y}_{32}$	0.002^{*}	0.000	0.002^{*}	0.000
$\delta_k \sim N(0, 0.05)$	$\bar{y}_{21}$	-0.002	0.006^{*}	0.000	0.004
	$\bar{y}_{22}$	0.002	0.000	0.000	0.000
	$\bar{y}_{31}$	-0.003	0.005	-0.002	0.003
	$\bar{y}_{32}$	0.001	0.000	0.001	0.000

Test of bias
*** significant at 0.1% level, ** significant at 1% level, * significant at 5% level

Table 5.2 *Approximate lower and upper 95% confidence limits for the variance of the estimated means and approximate confidence limits for the mean of the estimated variances when Mak & Li's estimator is used (all figures* $\times 10^3$).

			Estimator: Mak & Li				
Case	*Measurement*	*Parameter*	S_L^2	$S^2(\hat{\bar{y}}_{id})$	S_U^2	$\bar{\hat{V}}(\hat{\bar{y}}_{id})$	$\pm 1.96\sqrt{S^2(\bar{\hat{V}})}$
A)	Pr(true)=0.9	$\bar{y}_{11}$	7.38	8.04^{+++}	8.80	$3.96^{\times\times}$	0.178
		$\bar{y}_{12}$	1.58	1.72	1.88	$1.14^{\times\times\times}$	0.021
	$\delta_k \sim N(0, 0.1y_k)$	$\bar{y}_{21}$	5.77	6.28^{+++}	6.88	$5.30^{\times\times}$	0.144
		$\bar{y}_{22}$	0.820	0.893^{+}	0.978	$1.06^{\times\times\times}$	0.019
		$\bar{y}_{31}$	5.77	6.28^{+++}	6.88	$5.32^{\times\times}$	0.141
		$\bar{y}_{32}$	0.797	0.868	0.950	$1.06^{\times\times}$	0.017
	$\delta_k \sim N(0, 0.05)$	$\bar{y}_{21}$	5.71	6.22^{+++}	6.81	$5.25^{\times\times}$	0.165
		$\bar{y}_{22}$	0.823	0.896^{+}	0.980	$1.04^{\times\times}$	0.060
		$\bar{y}_{31}$	5.70	6.20^{+++}	6.79	$5.29^{\times\times}$	0.162
		$\bar{y}_{32}$	0.799	0.870	0.950	$1.04^{\times\times\times}$	0.024
B)	Pr(true)=0.9	$\bar{y}_{11}$	7.34	7.99^{+++}	8.75	$4.14^{\times\times\times}$	0.180
		$\bar{y}_{12}$	1.56	1.70	1.86	$1.02^{\times\times\times}$	0.067
	$\delta_k \sim N(0, 0.1y_k)$	$\bar{y}_{21}$	5.50	5.99^{+}	6.56	$5.07^{\times\times}$	0.146
		$\bar{y}_{22}$	0.889	0.968^{+}	1.06	1.04	0.019
		$\bar{y}_{31}$	5.44	5.92^{++}	6.48	$5.09^{\times\times}$	0.141
		$\bar{y}_{32}$	0.869	0.946^{+}	1.04	1.03	0.017
	$\delta_k \sim N(0, 0.05)$	$\bar{y}_{21}$	5.63	6.13^{+}	6.71	$5.16^{\times\times}$	0.161
		$\bar{y}_{22}$	0.903	0.983^{+}	1.08	1.06	0.024
		$\bar{y}_{31}$	5.55	6.05^{++}	6.62	$5.20^{\times\times}$	0.157
		$\bar{y}_{32}$	0.881	0.959^{+}	1.05	$1.05^{\times}$	0.021

Test of $\bar{\hat{V}}(\hat{\bar{y}}_{id}) - S^2(\hat{\bar{y}}_{id})$

$\times\times\times$ significant at 0.1% level, $\times\times$ significant at 1% level, $\times$ significant at 5% level

Test of $S^2_{m\text{GREG}}$-$S^2_{\text{M\&L}}$

+++ significant at 0.1% level, ++ significant at 1% level, + significant at 5% level

Table 5.3 *Approximate lower and upper 95% confidence limits for the variance of the estimated means and approximate confidence limits for the mean of the estimated variances when the mGREG estimator is used (all figures* $\times 10^3$).

			Estimator: mGREG				
Case	*Measurement*	*Parameter*	S_L^2	$S^2(\hat{\bar{y}}_{id})$	S_U^2	$\bar{\hat{V}}(\hat{\bar{y}}_{id})$	$\pm 1.96\sqrt{S^2(\bar{\hat{V}})}$
A)	Pr(true)=0.9	$\bar{y}_{11}$	9.29	10.1^{+++}	11.1	$11.7^{\times\times\times}$	0.405
		$\bar{y}_{12}$	1.60	1.74	1.90	1.83	0.026
	$\delta_k \sim N(0, 0.1y_k)$	$\bar{y}_{21}$	8.28	9.02^{+++}	9.87	9.48	0.400
		$\bar{y}_{22}$	0.865	0.942^{+}	1.03	1.01	0.015
		$\bar{y}_{31}$	8.15	8.87^{+++}	9.71	9.32	0.391
		$\bar{y}_{32}$	0.845	0.920	1.01	0.993	0.015
	$\delta_k \sim N(0, 0.05)$	$\bar{y}_{21}$	8.02	8.73^{+++}	9.56	8.86	0.296
		$\bar{y}_{22}$	0.849	0.924^{+}	1.01	0.972	0.014
		$\bar{y}_{31}$	7.91	8.61^{+++}	9.43	8.70	0.293
		$\bar{y}_{32}$	0.831	0.905	0.991	0.961	0.013
B)	Pr(true)=0.9	$\bar{y}_{11}$	10.6	11.5^{+++}	12.6	$14.8^{\times\times\times}$	0.627
		$\bar{y}_{12}$	1.54	1.68	1.84	1.78	0.026
	$\delta_k \sim N(0, 0.1y_k)$	$\bar{y}_{21}$	6.06	6.60^{+}	7.22	6.99	0.219
		$\bar{y}_{22}$	0.851	0.927^{+}	1.01	1.00	0.013
		$\bar{y}_{31}$	5.93	6.45^{++}	7.06	6.86	0.217
		$\bar{y}_{32}$	0.834	0.908^{+}	0.994	.988	0.013
	$\delta_k \sim N(0, 0.05)$	$\bar{y}_{21}$	6.40	6.97^{+}	7.63	7.14	0.205
		$\bar{y}_{22}$	0.862	0.938^{+}	1.03	$0.999^{\times}$	0.014
		$\bar{y}_{31}$	6.26	6.82^{++}	7.46	7.01	0.202
		$\bar{y}_{32}$	0.847	0.922^{+}	1.01	0.988	0.014

Test of $\bar{\hat{V}}(\hat{\bar{y}}_{id}) - S^2(\hat{\bar{y}}_{id})$

$\times\times\times$ significant at 0.1% level, $\times\times$ significant at 1% level, $\times$ significant at 5% level

Test of $S^2{}_{m\text{GREG}} - S^2{}_{\text{M\&L}}$

+++ significant at 0.1% level, ++ significant at 1% level, + significant at 5% level

Table 5.4 *Observed coverage rates and coefficients of variation when Mak & Li's and the mGREG estimators are used.*

			Estimator			
			Mak & Li		*mGREG*	
Case	*Measurement*	*Parameter*	*CI95%*	*cve%*	*CI95%*	*cve%*
A)	Pr(true)=0.9	$\bar{y}_{11}$	74.6***	11.2	96.6*	12.6
		$\bar{y}_{12}$	87.9***	10.4	95.8	10.4
	δ_k~$N(0, 0.1y_k)$	$\bar{y}_{21}$	90.7***	9.91	95.0	11.9
		$\bar{y}_{22}$	95.6	7.47	96.1	7.67
		$\bar{y}_{31}$	90.7***	9.91	94.9	11.8
		$\bar{y}_{32}$	96.3	7.37	96.0	7.58
	δ_k~$N(0, 0.05)$	$\bar{y}_{21}$	90.8***	9.86	94.6	11.7
		$\bar{y}_{22}$	95.2	7.48	95.7	7.60
		$\bar{y}_{31}$	90.7***	9.84	94.3	11.6
		$\bar{y}_{32}$	95.9	7.37	95.6	7.52
B)	Pr(true)=0.9	$\bar{y}_{11}$	75.4***	11.2	96.2	13.4
		$\bar{y}_{12}$	85.5***	10.3	95.5	10.2
	δ_k~$N(0, 0.1y_k)$	$\bar{y}_{21}$	90.5***	9.67	95.5	10.2
		$\bar{y}_{22}$	93.8	7.78	95.4	7.61
		$\bar{y}_{31}$	91.1***	9.62	95.2	10.0
		$\bar{y}_{32}$	95.2	7.69	95.4	7.53
	δ_k~$N(0, 0.05)$	$\bar{y}_{21}$	90.2***	9.79	94.5	10.4
		$\bar{y}_{22}$	93.1**	7.84	95.8	7.66
		$\bar{y}_{31}$	90.8***	9.72	94.2	10.3
		$\bar{y}_{32}$	94.1	7.74	95.9	7.59

*** significant at 0.1% level, ** significant at 1% level, * significant at 5% level

Table 5.5 *mGREG estimates of* $\bar{y}_{1d}$ *in domains* 1 *and* 2 *when the same auxiliary variable (Aux var) is used in the numerator and the denominator.*

		Domain 1				*Domain 2*			
Case	*Aux var*	*Bias*	S^2 $\times 10^3$	$\bar{\hat{V}}$ $\times 10^3$	*CI95 %*	*Bias*	S^2 $\times 10^3$	$\bar{\hat{V}}$ $\times 10^3$	*CI95 %*
A)	x_{dk}	0.002	8.23	7.86	90.8***	-0.003	2.84	3.03	95.3
	z_{dk}	0.003	8.08	7.79	91.1***	-0.001	2.81	3.00	95.6
	x_{dk}, z_{dk}	0.011***	10.1	11.7$^{\times\times\times}$	96.6*	0.001	1.74	1.83	95.8
B)	x_{dk}	0.003	8.07	7.72	91.2***	-0.004*	2.65	2.87	95.4
	z_{dk}	0.004	8.05	7.75	91.2***	-0.001	2.73	2.96	95.4
	x_{dk}, z_{dk}	0.013***	11.5	14.8$^{\times\times\times}$	96.2	0.001	1.68	1.78	95.5

Test of bias
*** significant at 0.1% level, ** significant at 1% level, * significant at 5% level

Test of $\bar{\hat{V}}(\hat{\bar{y}}_{id}) - S^2(\hat{\bar{y}}_{id})$
$\times\times\times$ significant at 0.1% level, $\times\times$ significant at 1% level, $\times$ significant at 5% level

Table 5.6 *mGREG estimates of* $\bar{y}_{11}$ *in domain* 1 *when the same auxiliary variable (Aux var) is used in the numerator and the denominator only when an inadmissible estimate is encountered.*

		Domain 1			
Case	*Aux var*	*Bias*	S^2 $\times 10^3$	$\bar{\hat{V}}$ $\times 10^3$	*CI95 %*
A)	x_{dk}	0.001	8.51	10.4$^{\times\times\times}$	95.1
	z_{dk}	0.001	8.51	10.5$^{\times\times\times}$	95.1
	x_{dk}, z_{dk}	0.011***	10.1	11.7$^{\times\times\times}$	96.6*
B)	x_{dk}	0.000	9.50	12.5$^{\times\times\times}$	94.7
	z_{dk}	0.000	9.49	12.5$^{\times\times\times}$	94.7
	x_{dk}, z_{dk}	0.013***	11.5	14.8$^{\times\times\times}$	96.2

Test of bias
*** significant at 0.1% level, ** significant at 1% level, * significant at 5% level

Test of $\bar{\hat{V}}(\hat{\bar{y}}_{id}) - S^2(\hat{\bar{y}}_{id})$
$\times\times\times$ significant at 0.1% level, $\times\times$ significant at 1% level, $\times$ significant at 5% level

Fig 5.1-5.8 *Distribution of* 1,000 *estimated means,* $\hat{\bar{y}}_{1d}$ *and variances,* $\hat{V}(\hat{\bar{y}}_{1d})$, *d=1, 2, Case=A,* Pr(*true*)=0.9 *when Mak & Li and the mGREG estimators are used.*

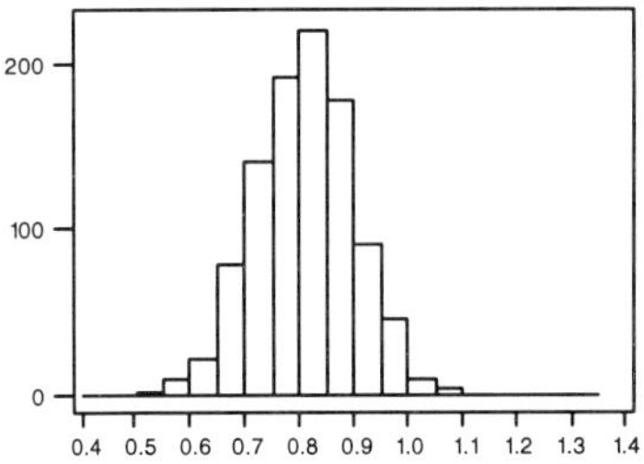

Fig 5.1 *Mak & Li,* $\hat{\bar{y}}_{11}$

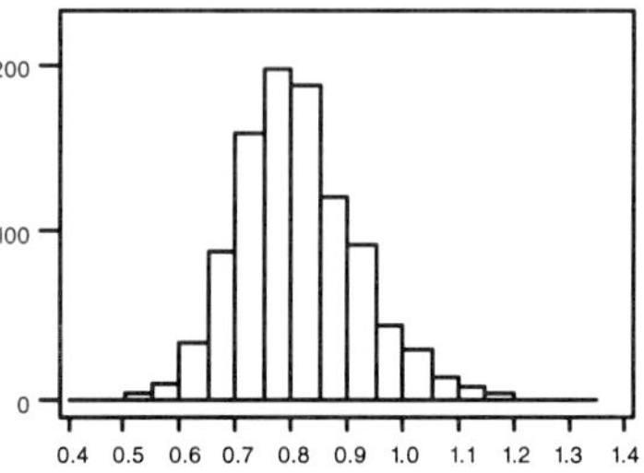

Fig 5.2 *mGREG,* $\hat{\bar{y}}_{11}$

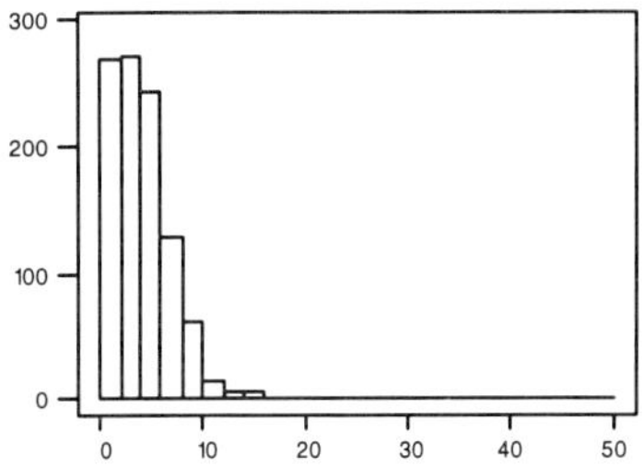

Fig 5.3 *Mak & Li,* $\hat{V}(\hat{\bar{y}}_{11}) \times 10^3$

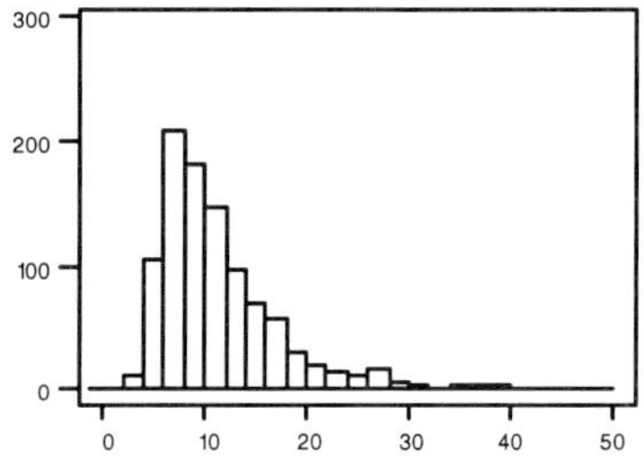

Fig 5.4 *mGREG,* $\hat{V}(\hat{\bar{y}}_{11}) \times 10^3$

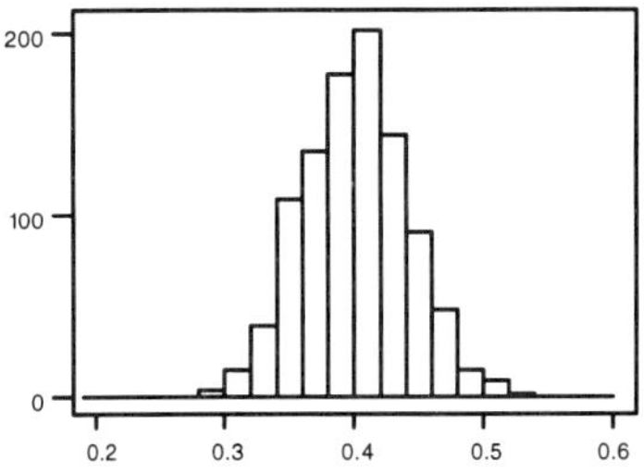

Fig 5.5 *Mak & Li,* $\hat{\bar{y}}_{12}$

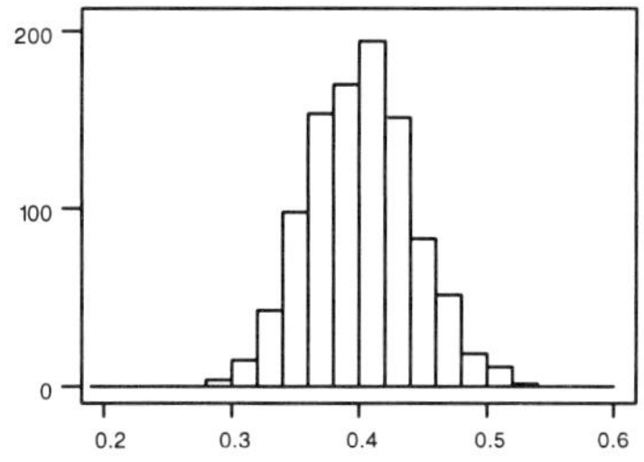

Fig 5.6 *mGREG,* $\hat{\bar{y}}_{12}$

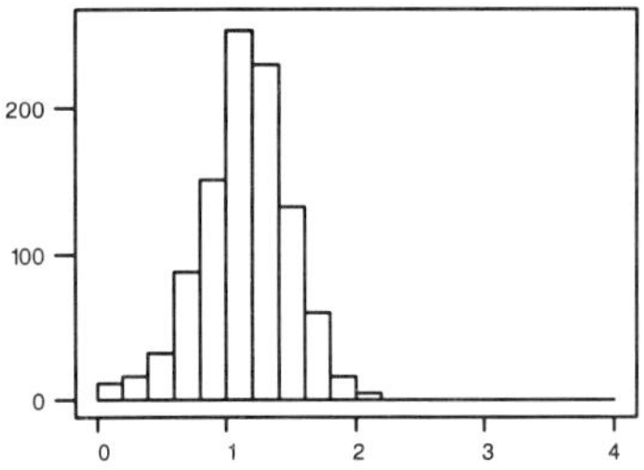

Fig 5.7 *Mak & Li,* $\hat{V}(\hat{\bar{y}}_{12}) \times 10^3$

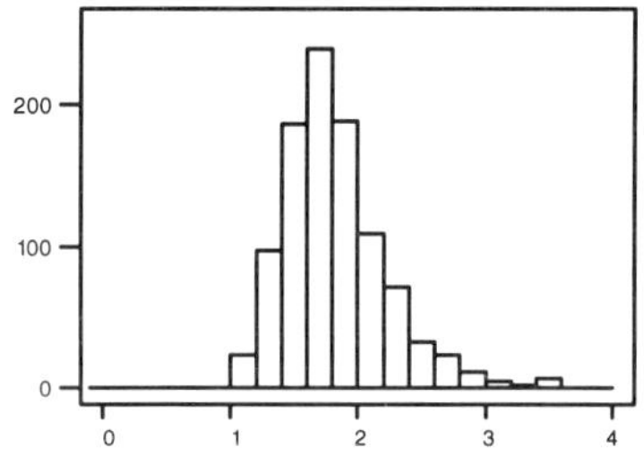

Fig 5.8 *mGREG,* $\hat{V}(\hat{\bar{y}}_{12}) \times 10^3$

Fig 5.9-5.16 *Distribution of* 1,000 *estimated means,* $\hat{\bar{y}}_{2d}$ *and variances,* $\hat{V}(\hat{\bar{y}}_{2d})$, *d*=1, 2, *Case=A,* $\delta_k \sim N(0,\ 0.1y_k)$ *when Mak & Li and the mGREG estimators are used.*

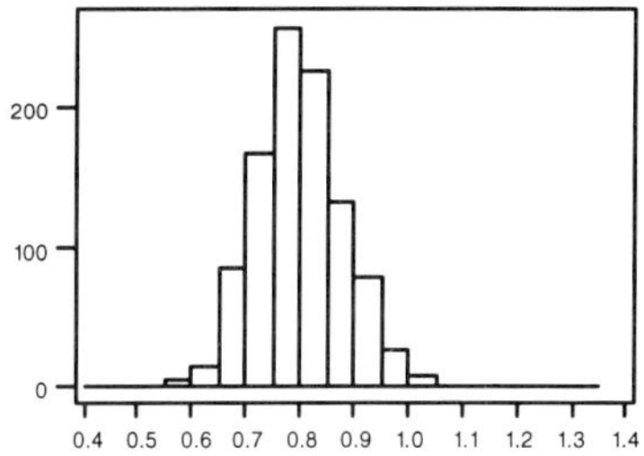

Fig 5.9 *Mak & Li,* $\hat{\bar{y}}_{21}$

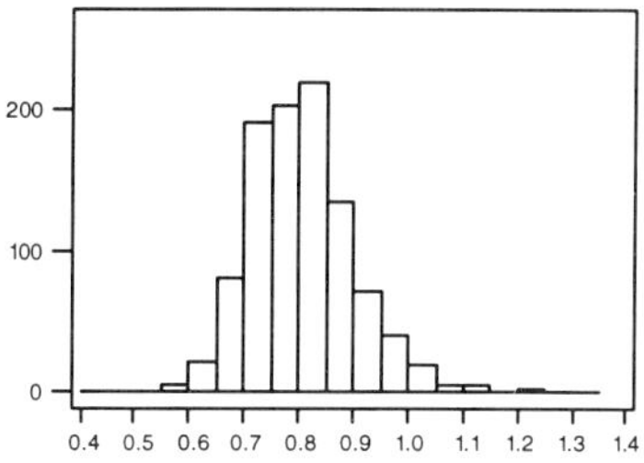

Fig 5.10 *mGREG,* $\hat{\bar{y}}_{21}$

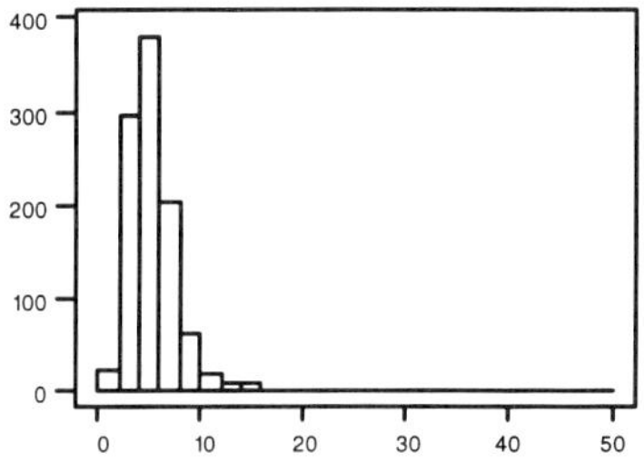

Fig 5.11 *Mak & Li,* $\hat{V}(\hat{\bar{y}}_{21}) \times 10^3$

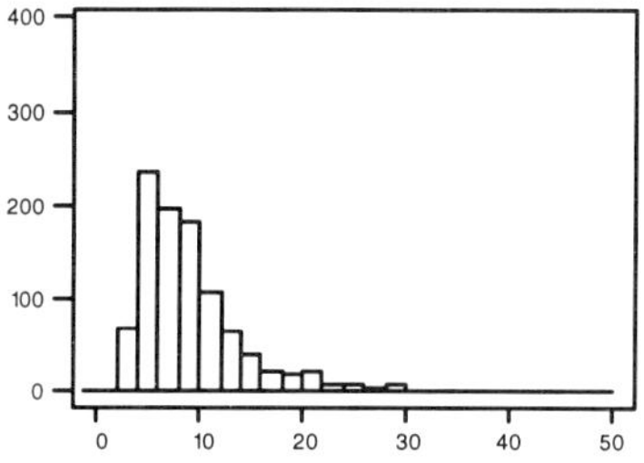

Fig 5.12 *mGREG,* $\hat{V}(\hat{\bar{y}}_{21}) \times 10^3$

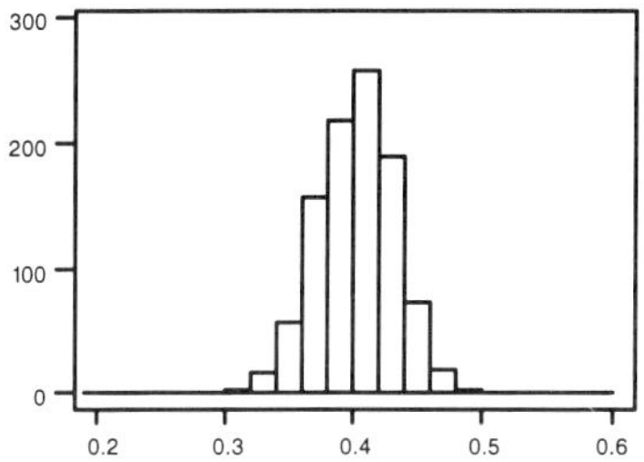

Fig 5.13 *Mak & Li,* $\hat{\bar{y}}_{22}$

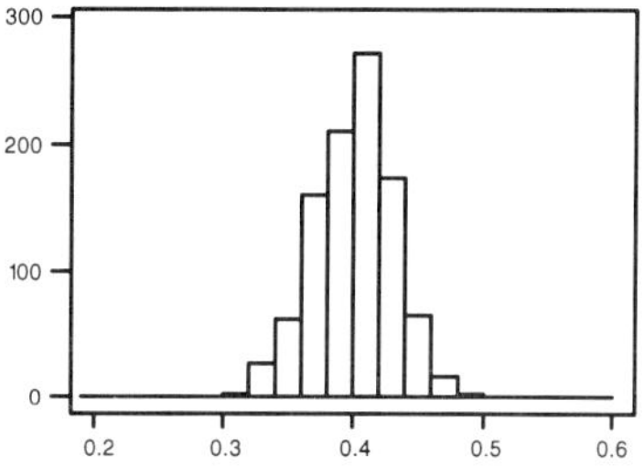

Fig 5.14 *mGREG,* $\hat{\bar{y}}_{22}$

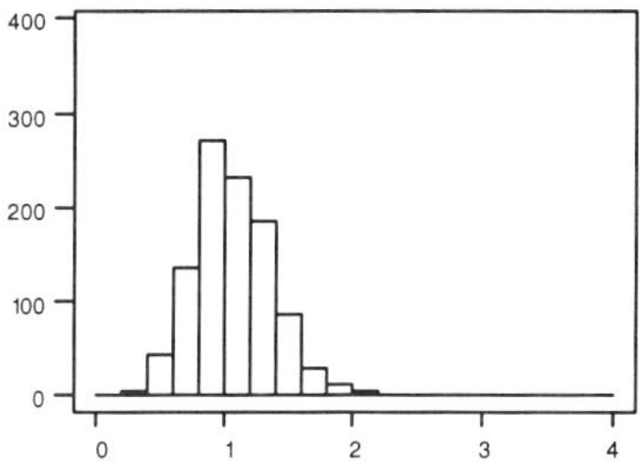

Fig 5.15 *Mak & Li,* $\hat{V}(\hat{\bar{y}}_{22}) \times 10^3$

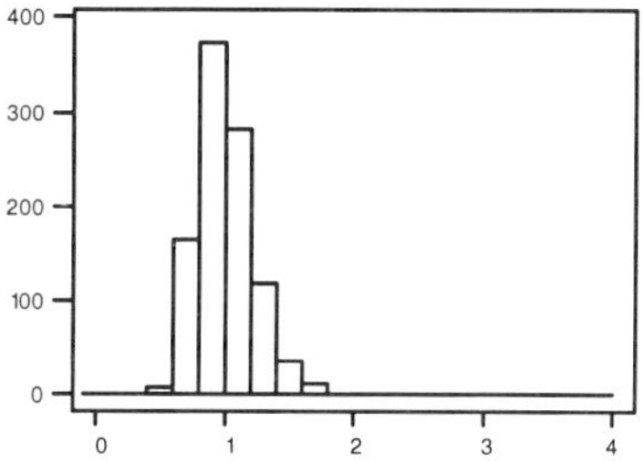

Fig 5.16 *mGREG,* $\hat{V}(\hat{\bar{y}}_{22}) \times 10^3$

Fig 5.17-5.24 *Distribution of* 1,000 *estimated means,* $\hat{\bar{y}}_{3d}$ *and variances,* $\hat{V}(\hat{\bar{y}}_{3d})$, *d*=1, 2, *Case*=A, $\delta_k \sim N(0,\ 0.1y_k)$ *when Mak & Li and the mGREG estimators are used.*

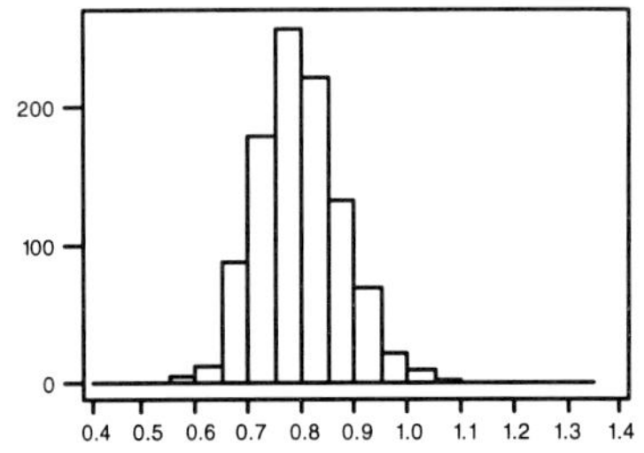

Fig 5.17 *Mak & Li,* $\hat{\bar{y}}_{31}$

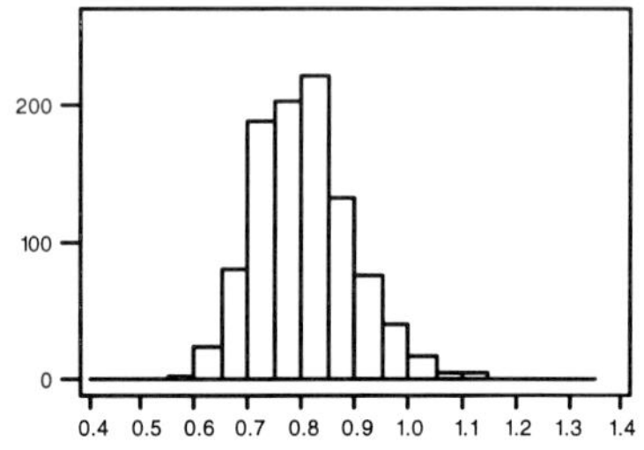

Fig 5.18 *mGREG,* $\hat{\bar{y}}_{31}$

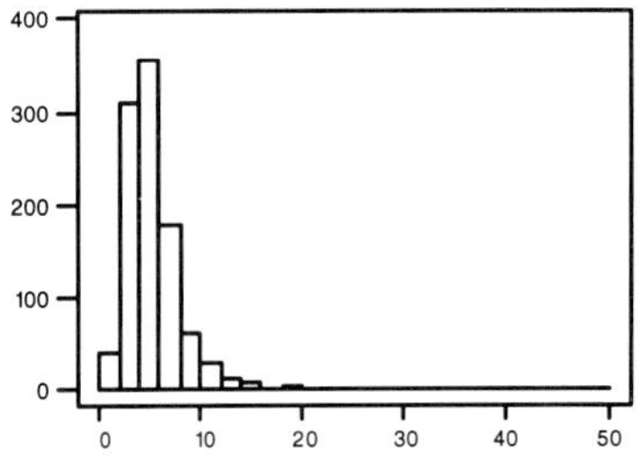

Fig 5.19 *Mak & Li,* $\hat{V}(\hat{\bar{y}}_{31})\times 10^3$

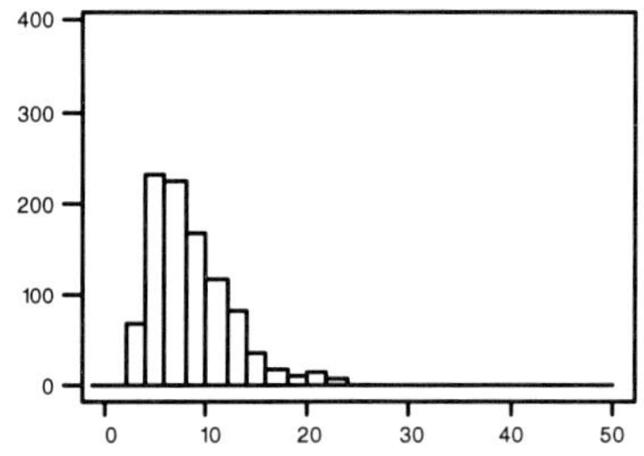

Fig 5.20 *mGREG,* $\hat{V}(\hat{\bar{y}}_{31})\times 10^3$

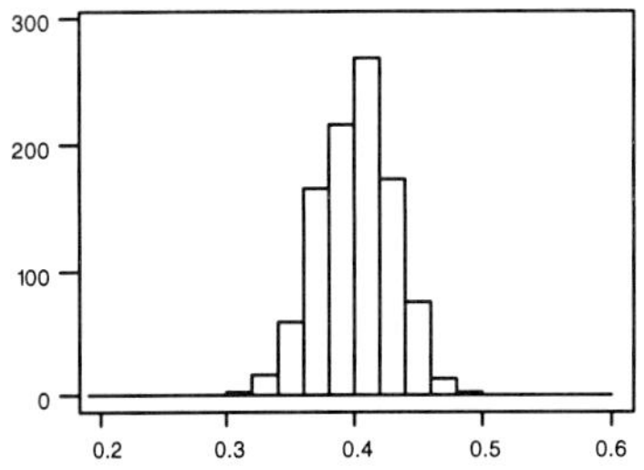

Fig 5.21 *Mak & Li,* $\hat{\bar{y}}_{32}$

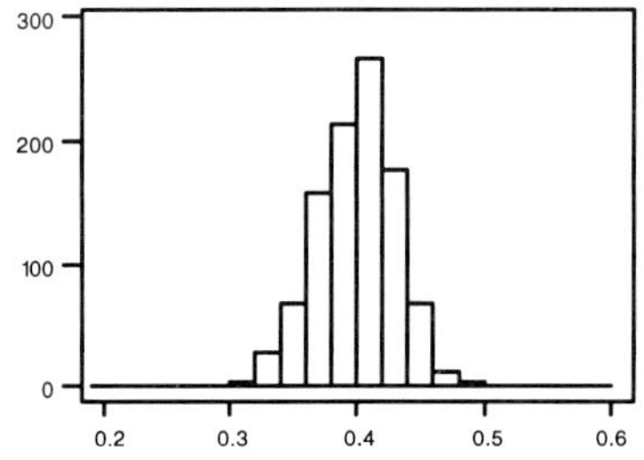

Fig 5.22 *mGREG,* $\hat{\bar{y}}_{32}$

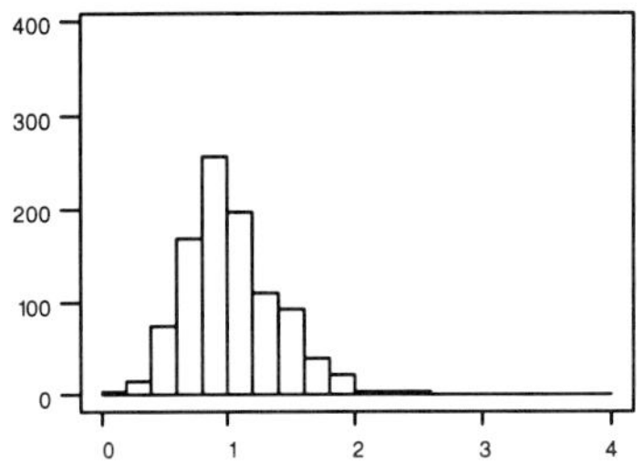

Fig 5.23 *Mak & Li,* $\hat{V}(\hat{\bar{y}}_{32})\times 10^3$

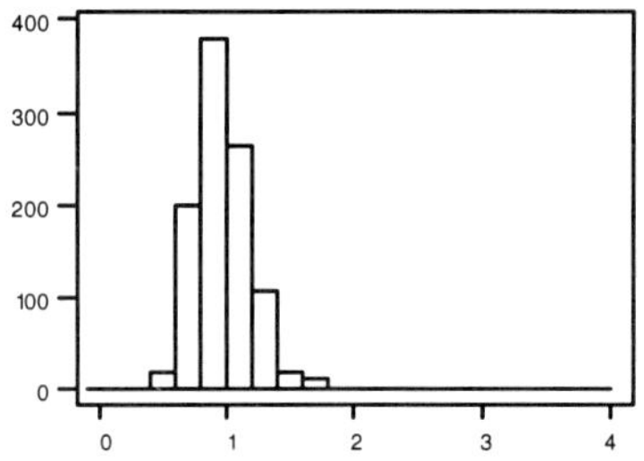

Fig 5.24 *mGREG,* $\hat{V}(\hat{\bar{y}}_{32})\times 10^3$

Fig 5.25-5.32 *Distribution of* 1,000 *estimated means,* $\hat{\bar{y}}_{1d}$ *and variances,* $\hat{V}(\hat{\bar{y}}_{1d})$, *d*=1, 2, *Case=B*, Pr(*true*)=0.9 *when Mak & Li and the mGREG estimators are used.*

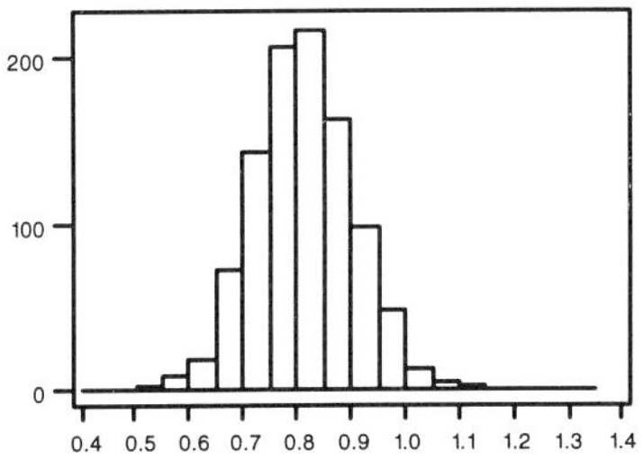

Fig 5.25 *Mak & Li,* $\hat{\bar{y}}_{11}$

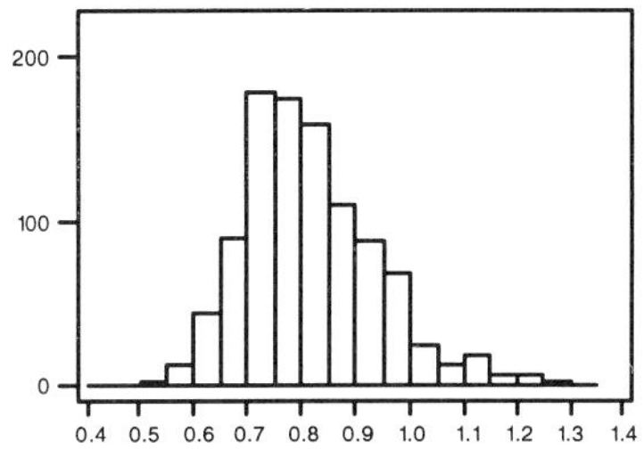

Fig 5.26 *mGREG,* $\hat{\bar{y}}_{11}$

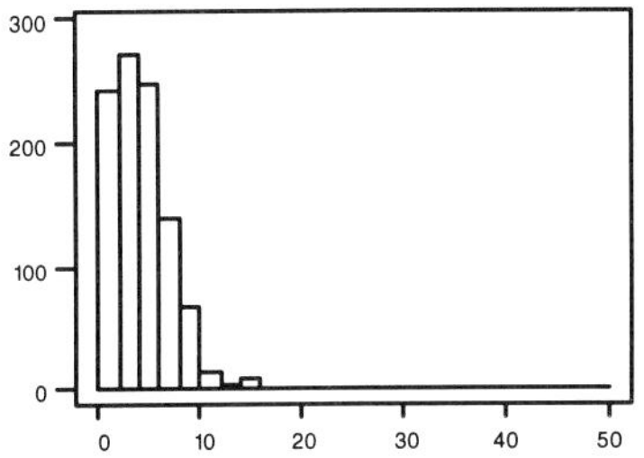

Fig 5.27 *Mak & Li,* $\hat{V}(\hat{\bar{y}}_{11})\times 10^3$

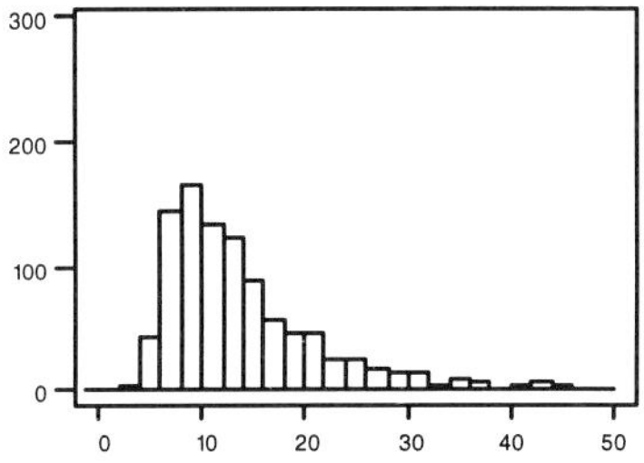

Fig 5.28 *mGREG,* $\hat{V}(\hat{\bar{y}}_{11})\times 10^3$

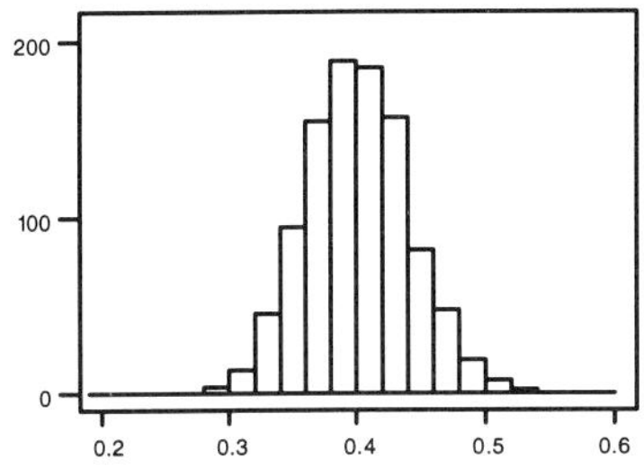

Fig 5.29 *Mak & Li,* $\hat{\bar{y}}_{12}$

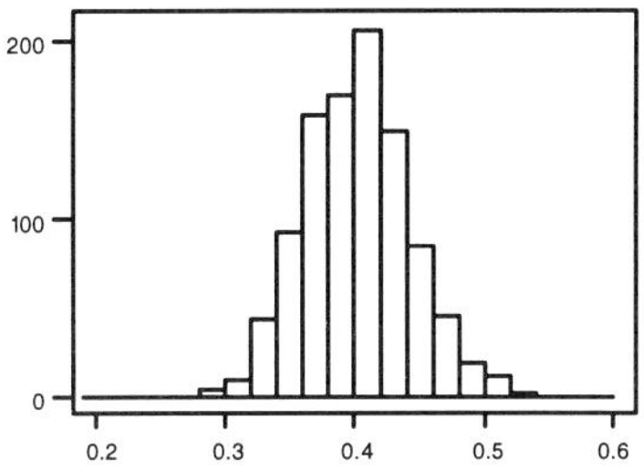

Fig 5.30 *mGREG,* $\hat{\bar{y}}_{12}$

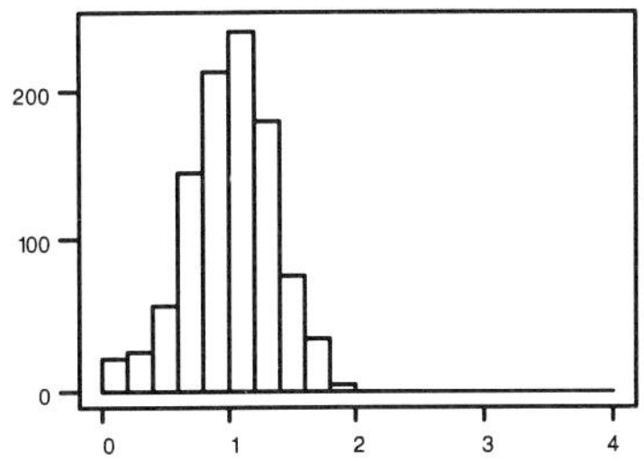

Fig 5.31 *Mak & Li,* $\hat{V}(\hat{\bar{y}}_{12})\times 10^3$

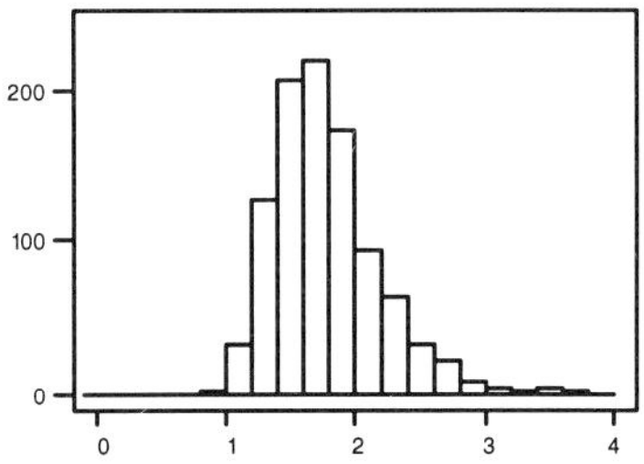

Fig 5.32 *mGREG,* $\hat{V}(\hat{\bar{y}}_{12})\times 10^3$

6 – 340198. Andersson, Claes: On the Use...

Fig 5.33-5.40 *Distribution of* 1,000 *estimated means,* $\hat{\bar{y}}_{2d}$ *and variances,* $\hat{V}(\hat{\bar{y}}_{2d})$, d=1, 2, *Case=B*, $\delta_k \sim N(0,\ 0.1y_k)$ *when Mak & Li and the mGREG estimators are used.*

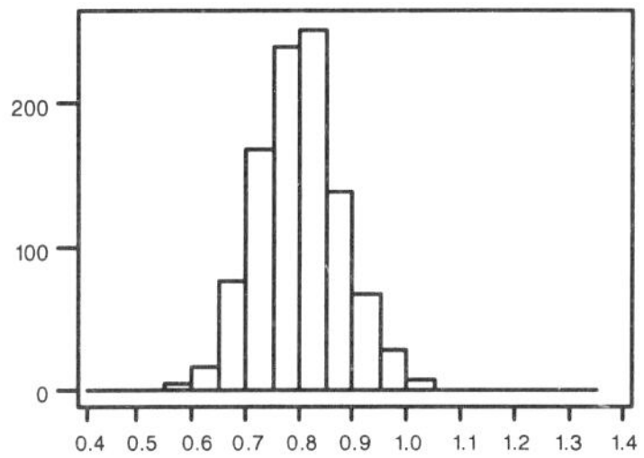

Fig 5.33 *Mak & Li,* $\hat{\bar{y}}_{21}$

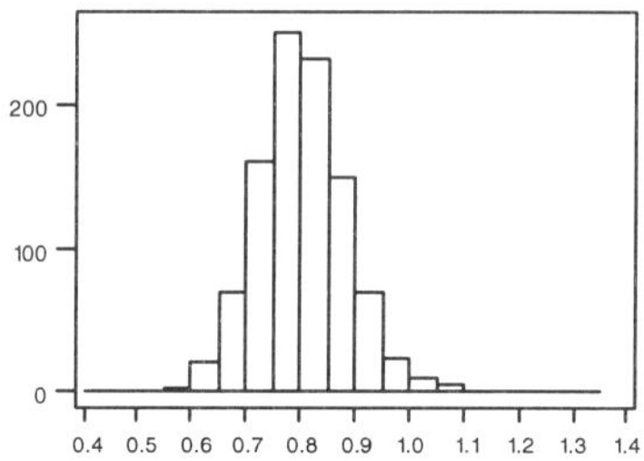

Fig 5.34 *mGREG,* $\hat{\bar{y}}_{21}$

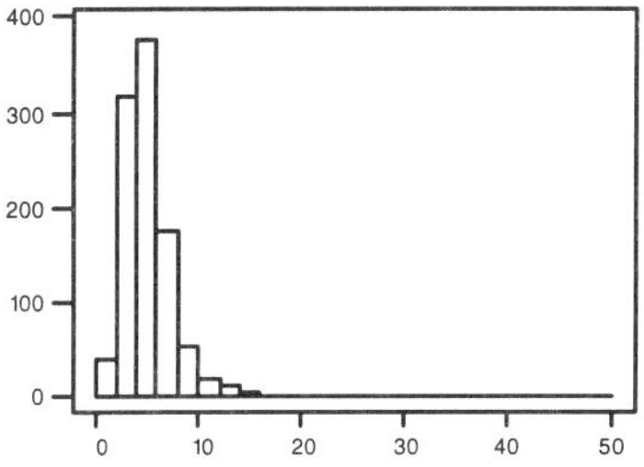

Fig 5.35 *Mak & Li,* $\hat{V}(\hat{\bar{y}}_{21}) \times 10^3$

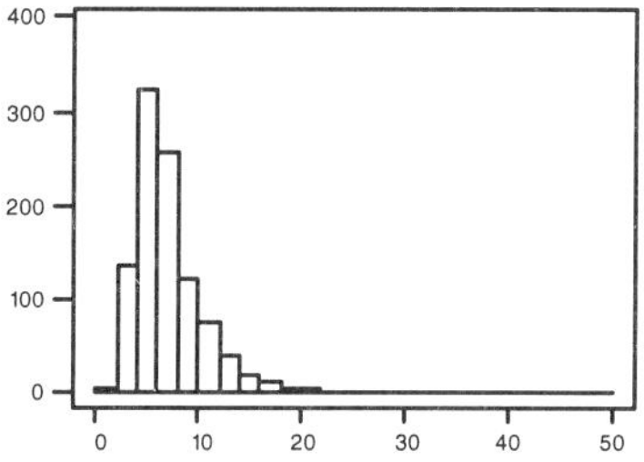

Fig 5.36 *mGREG,* $\hat{V}(\hat{\bar{y}}_{21}) \times 10^3$

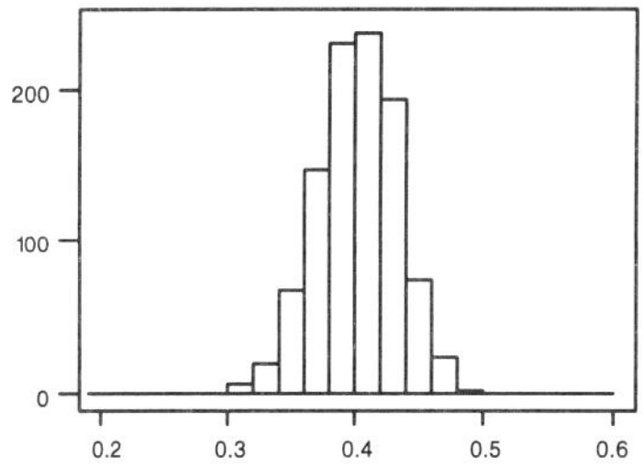

Fig 5.37 *Mak & Li,* $\hat{\bar{y}}_{22}$

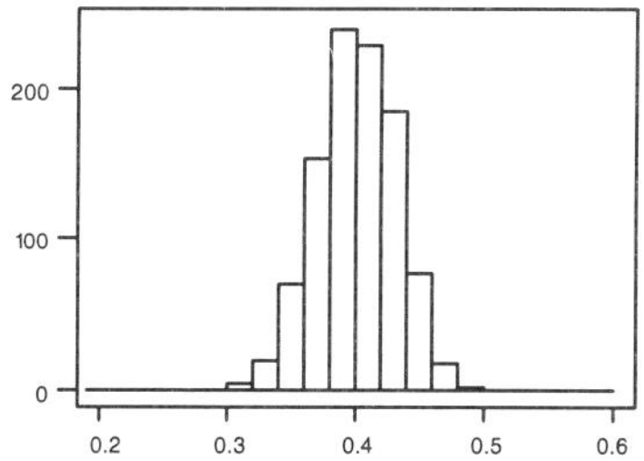

Fig 5.38 *mGREG,* $\hat{\bar{y}}_{22}$

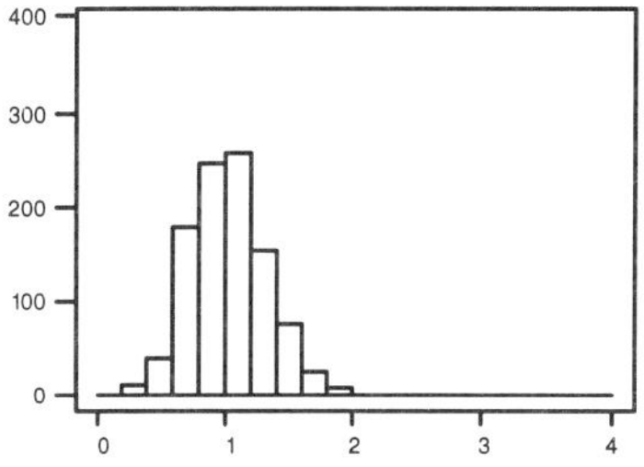

Fig 5.39 *Mak & Li,* $\hat{V}(\hat{\bar{y}}_{22}) \times 10^3$

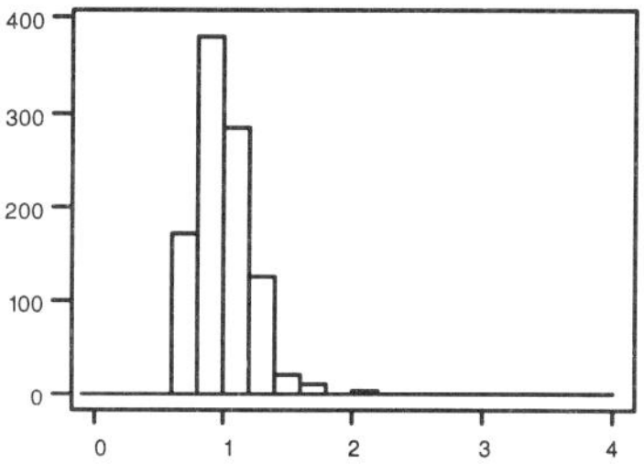

Fig 5.40 *mGREG,* $\hat{V}(\hat{\bar{y}}_{22}) \times 10^3$

Fig 5.41-5.48 *Distribution of* 1,000 *estimated means,* $\hat{\bar{y}}_{3d}$ *and variances,* $\hat{V}(\hat{\bar{y}}_{3d})$, d=1, 2, *Case=B,* $\delta_k \sim N(0,\ 0.1y_k)$ *when Mak & Li and the mGREG estimators are used.*

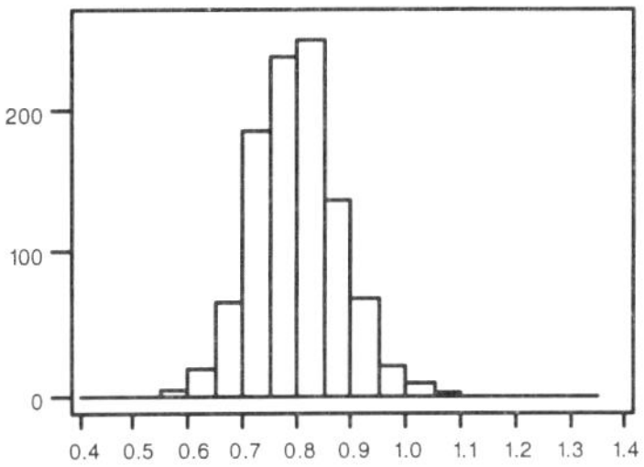

Fig 5.41 *Mak & Li,* $\hat{\bar{y}}_{31}$

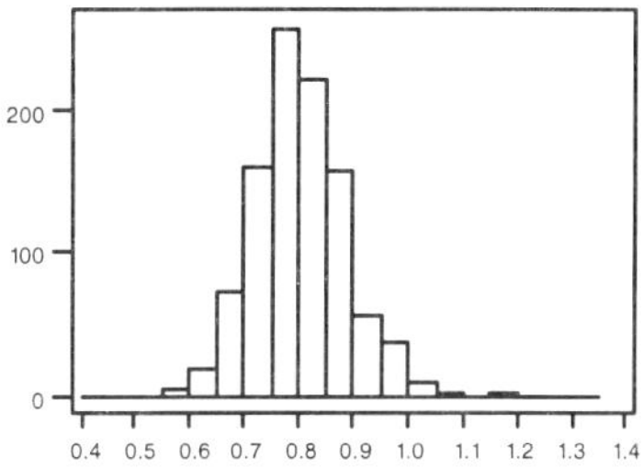

Fig 5.42 *mGREG,* $\hat{\bar{y}}_{31}$

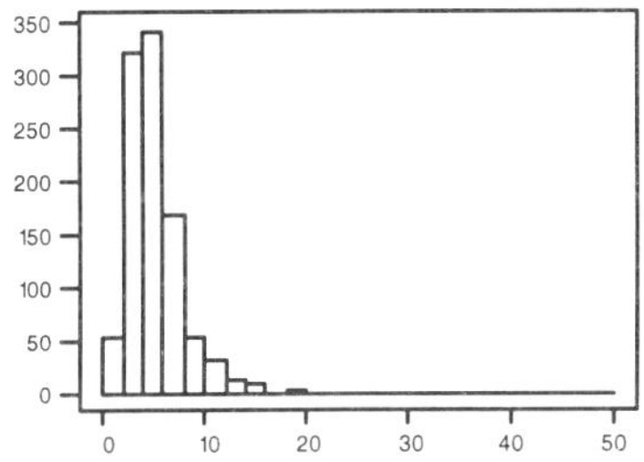

Fig 5.43 *Mak & Li,* $\hat{V}(\hat{\bar{y}}_{31}) \times 10^3$

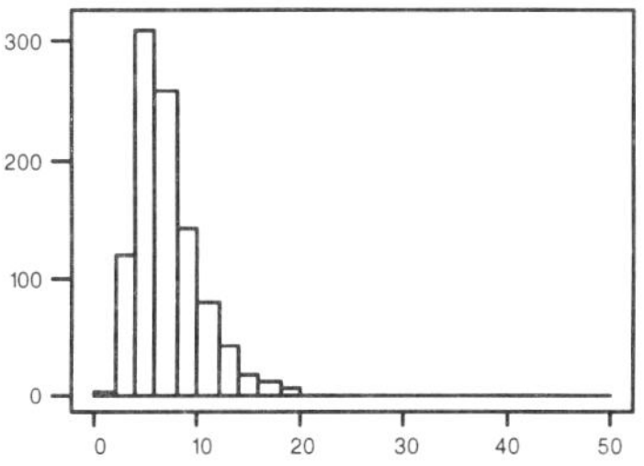

Fig 5.44 *mGREG,* $\hat{V}(\hat{\bar{y}}_{31}) \times 10^3$

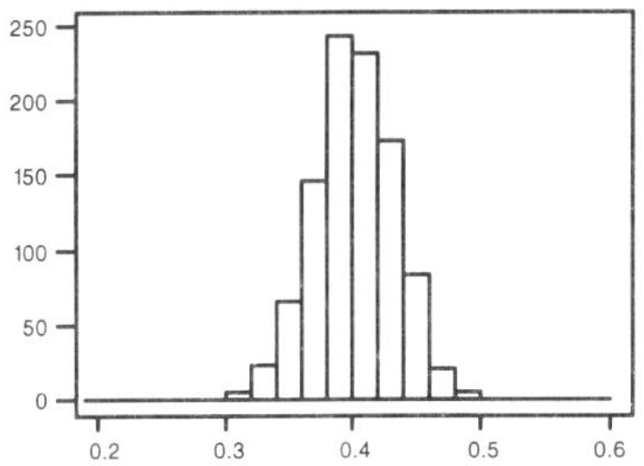

Fig 5.45 *Mak & Li,* $\hat{\bar{y}}_{32}$

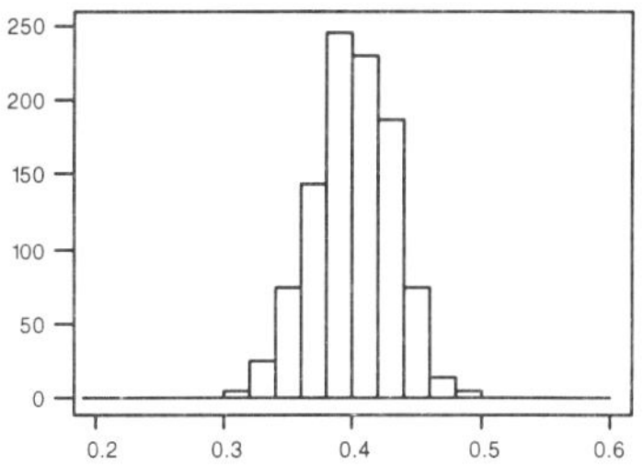

Fig 5.46 *mGREG,* $\hat{\bar{y}}_{32}$

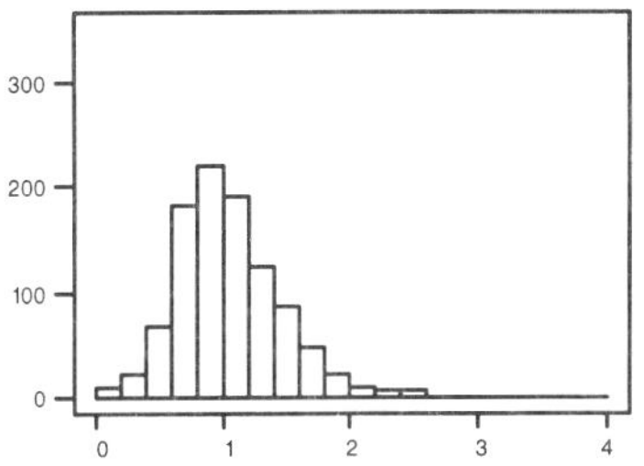

Fig 5.47 *Mak & Li,* $\hat{V}(\hat{\bar{y}}_{32}) \times 10^3$

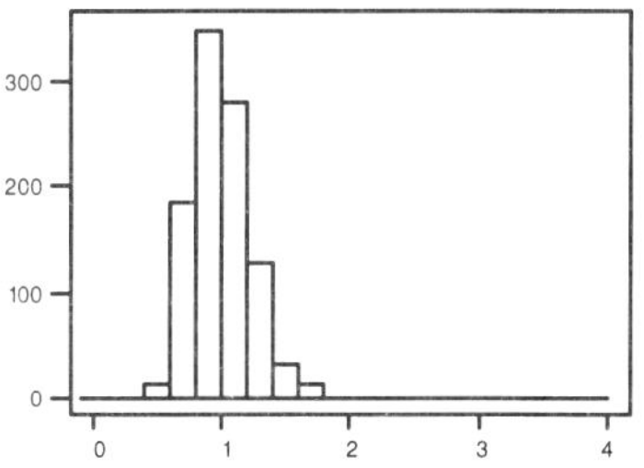

Fig 5.48 *mGREG,* $\hat{V}(\hat{\bar{y}}_{32}) \times 10^3$

Fig 5.49-5.56 *Distribution of* 1,000 *estimated means,* $\hat{\bar{y}}_{1d}$ *and variances,* $\hat{V}(\hat{\bar{y}}_{1d})$, *d*=1, 2, *Case=A*, Pr(*true*)=0.9, *mGREG and the same auxiliary variable* (*Aux var*) *in the numerator and the denominator.*

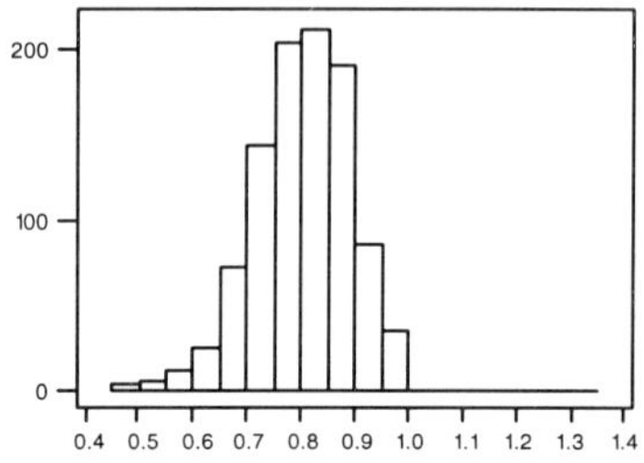

Fig 5.49 $\hat{\bar{y}}_{11}$, *Aux var=x*

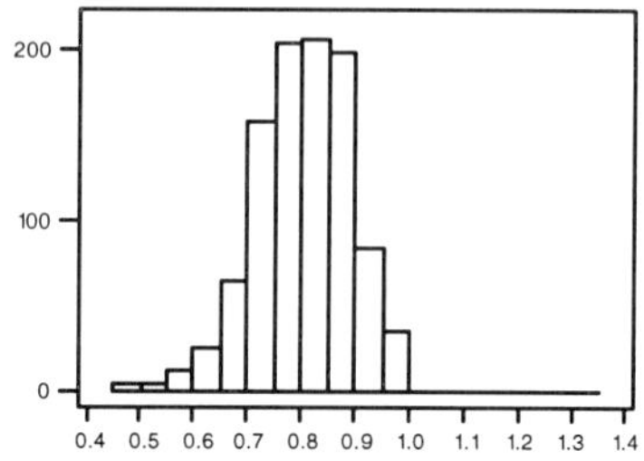

Fig 5.50 $\hat{\bar{y}}_{11}$, *Aux var=z*

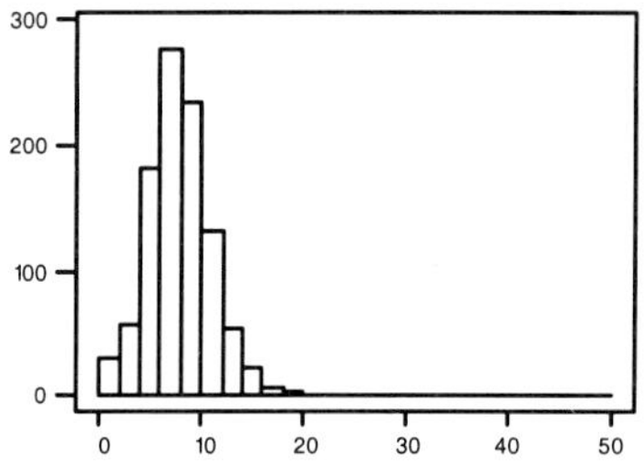

Fig 5.51 $\hat{V}(\hat{\bar{y}}_{11})\times 10^3$, *Aux var=x*

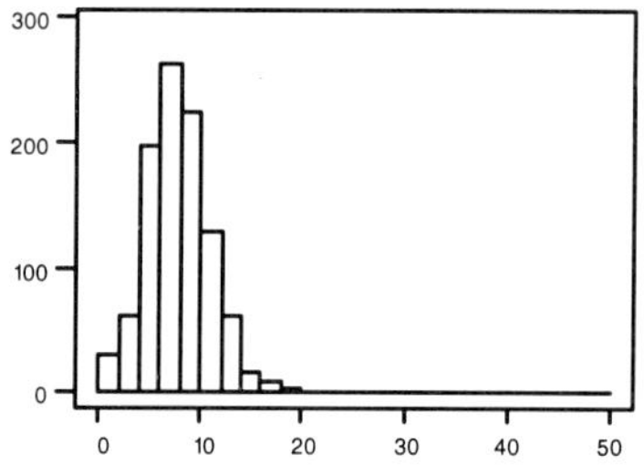

Fig 5.52 $\hat{V}(\hat{\bar{y}}_{11})\times 10^3$, *Aux var=z*

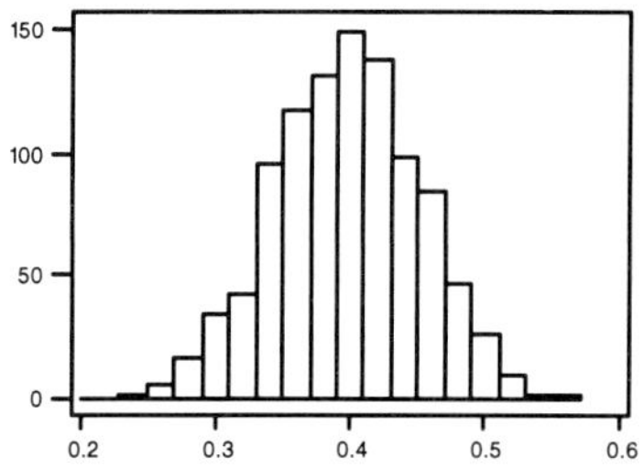

Fig 5.53 $\hat{\bar{y}}_{12}$, *Aux var=x*

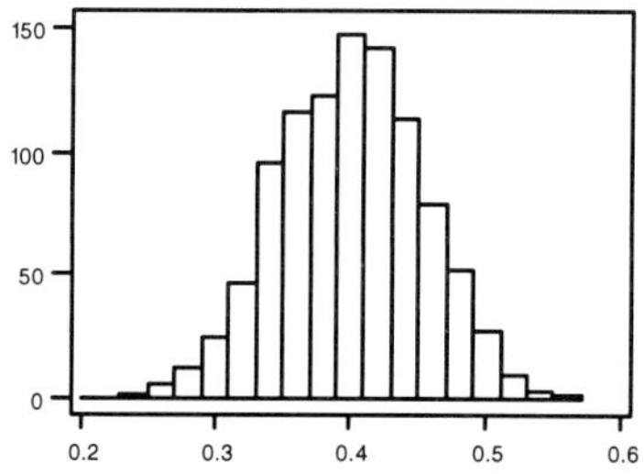

Fig 5.54 $\hat{\bar{y}}_{12}$, *Aux var=z*

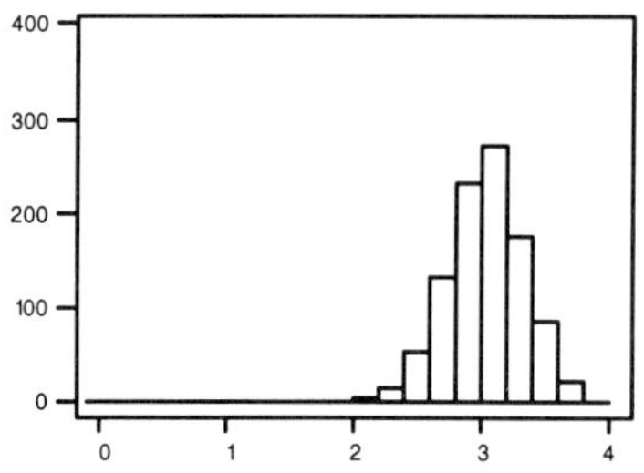

Fig 5.55 $\hat{V}(\hat{\bar{y}}_{12})\times 10^3$, *Aux var=x*

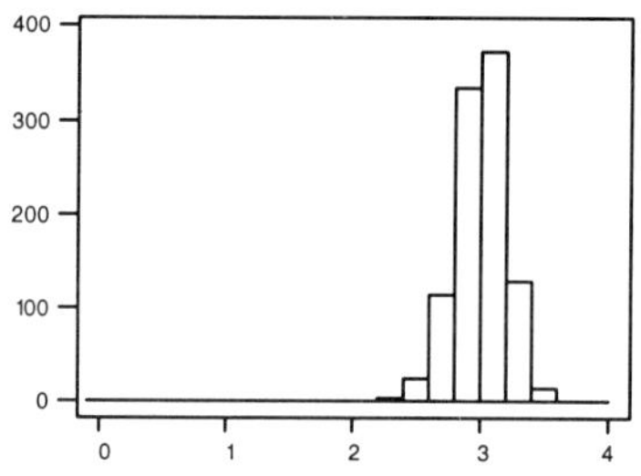

Fig 5.56 $\hat{V}(\hat{\bar{y}}_{12})\times 10^3$, *Aux var=z*

Fig 5.57-5.64 *Distribution of* 1,000 *estimated means,* $\hat{\bar{y}}_{1d}$ *and variances,* $\hat{V}(\hat{\bar{y}}_{1d})$, *d*=1, 2, *Case=B*, Pr(*true*)=0.9, *mGREG and the same auxiliary variable* (*Aux var*) *in the numerator and the denominator.*

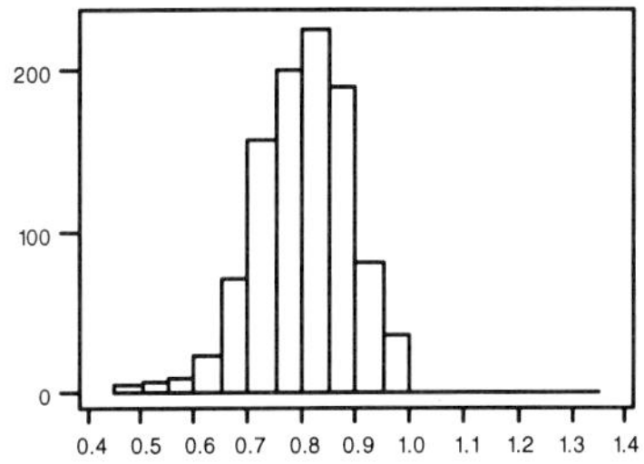

Fig 5.57 $\hat{\bar{y}}_{11}$, *Aux var=x*

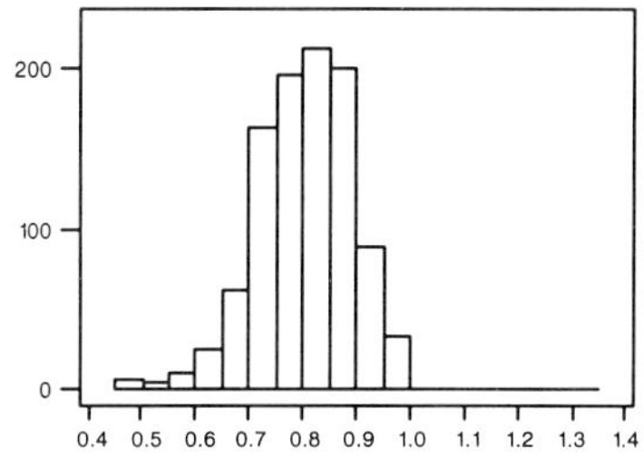

Fig 5.58 $\hat{\bar{y}}_{11}$, *Aux var=z*

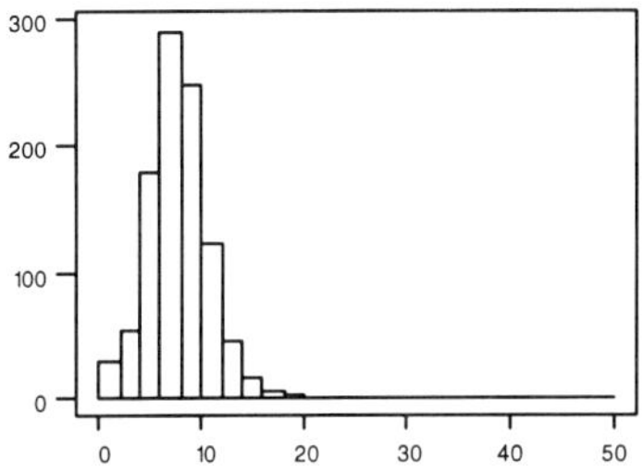

Fig 5.59 $\hat{V}(\hat{\bar{y}}_{11})\times 10^3$, *Aux var=x*

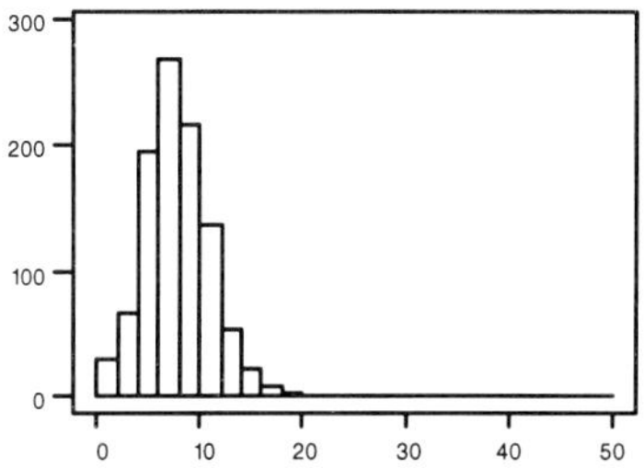

Fig 5.60 $\hat{V}(\hat{\bar{y}}_{11})\times 10^3$, *Aux var=z*

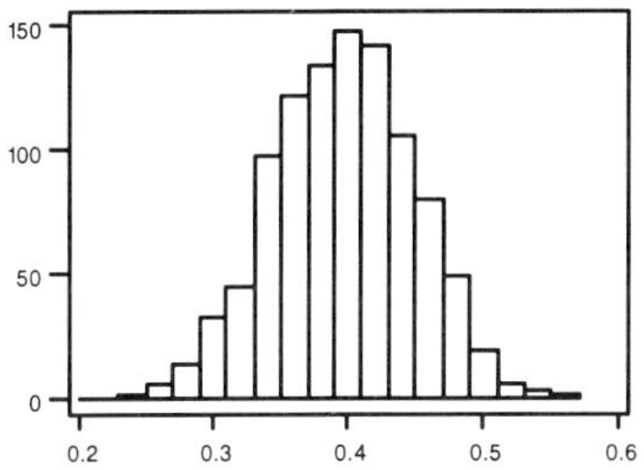

Fig 5.61 $\hat{\bar{y}}_{12}$, *Aux var=x*

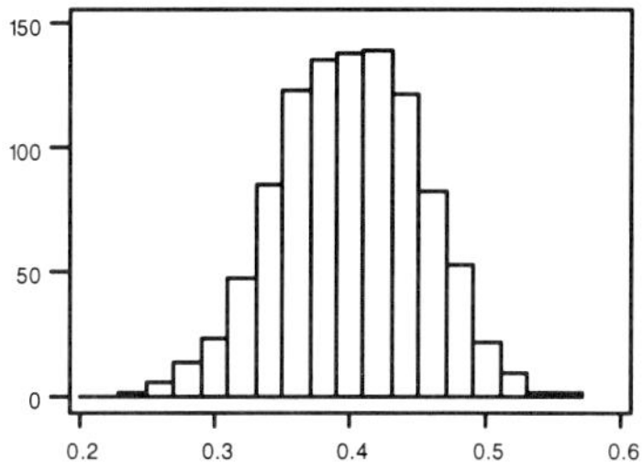

Fig 5.62 $\hat{\bar{y}}_{12}$, *Aux var=z*

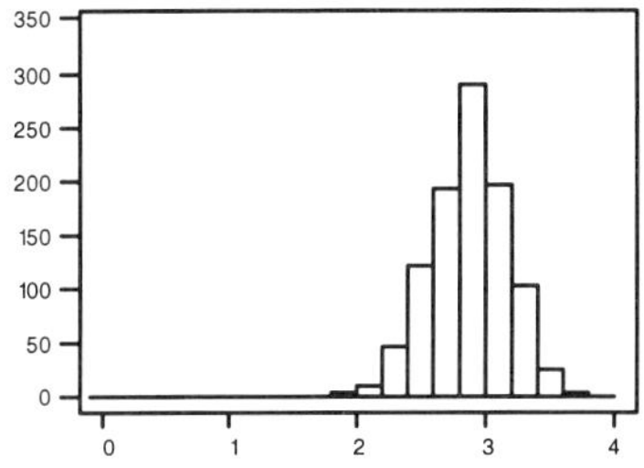

Fig 5.63 $\hat{V}(\hat{\bar{y}}_{12})\times 10^3$, *Aux var=x*

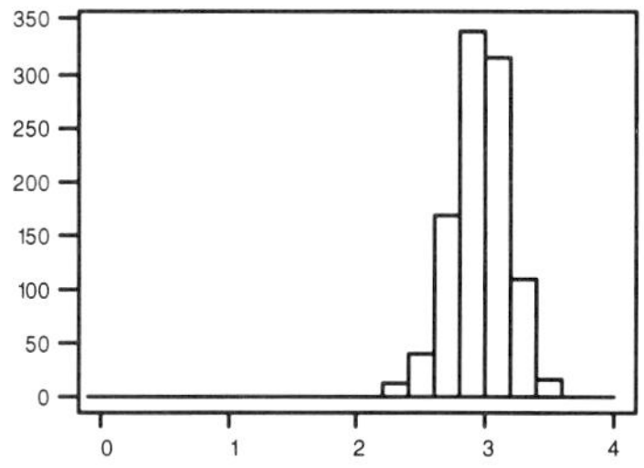

Fig 5.64 $\hat{V}(\hat{\bar{y}}_{12})\times 10^3$, *Aux var=z*

CHAPTER 6

Other Designs

6.1 Introduction

In the preceding chapters different estimators for the mean in a domain were proposed. The estimators proposed by Selén, Mak and Mak & Li cannot be used to estimate other parameters than a domain mean, and they cannot be used in situations where the samples are drawn according to other designs than *SI* or *SIR*.

The *m*GREG estimator can, however, be used under a general design, and it can also be used to form estimators for other parameters than the mean value. In this chapter the form of the *m*GREG estimator for *a total* under some common designs will be given. The limitation to a total is not very serious since most parameters of interest are functions of totals. The estimation of such functions and their variances can be obtained by using the method of Taylor linearization described in Appendix A.

Recall that the relation between the true y_{dk} and the error-prone $x_{d'k}$ values may be described by the model

$$\begin{cases} E_\xi(y_{dk}|\mathbf{x}_k) = \mathbf{x}'_k \boldsymbol{\beta}_d \\ V_\xi(y_{dk}|\mathbf{x}_k) = \sigma_d^2 x_{d'k} \quad k \to d' \end{cases} \tag{6.1}$$

The *m*GREG estimator then may be written

$$\hat{t}_{dr} = \hat{t}_d^* + (\hat{\mathbf{t}}_{ax} - \hat{\mathbf{t}}_x)' \mathbf{b}_d = \Sigma_s \frac{g_k y_{dk}}{\pi_k^*} \tag{6.2}$$

which simplifies to

$$\hat{t}_{dr} = \Sigma_s \frac{g_k y_{dk}}{\pi_k^*} = \sum_{d'=1}^{D} \frac{\hat{t}_{axd'}}{\hat{t}_{xd'}^*} \hat{t}_{dd'}^* \tag{6.3}$$

under the assumption of the variance pattern (6.1).

6.2 The *m*GREG Estimator in the Case of Stratification and *SI* Designs within Strata

Let the finite population $U=\{1, \ldots, k, \ldots, N\}$ be partitioned into H non-overlapping subpopulations called strata, and denoted $U_1, \ldots, U_h, \ldots, U_H$, where U_h consists of N_h units, and let $N = \sum_{h=1}^{H} N_h$.

The size of stratum h, N_h, is assumed to be known for every h. As denoted in section 1.2, the population is also assumed to be partitioned into D non-overlapping domains, $U_1, \ldots, U_d, \ldots, U_D$, of unknown sizes. The partitioning into strata and into domains may be done completely independent of each other. A population unit is supposed to belong to one and only one partition $U_{hd}=U_h \cap U_d$ of, usually unknown, size N_{hd}, due to the non-overlapping property. In each stratum h, a sample s_{ah} of size n_{ah} is taken according to the *SI* design, this design being called the *STSI* design.

Each unit in the sample $s_a=s_{a1} \cup \ldots \cup s_{ah} \ldots \cup s_{aH}$ is classified into one of D classes and assigned a value of the response variable y by the fallible procedure F. To secure correct classifications and measurements, a subsample s_h of size n_h is drawn from s_{ah} under the *SI* design, and the true procedure T is applied to every element in s_h. This is *one* way of taking the subsample, but it seems intuitively sound, since the original stratification usually serves some purpose and the subsample ought to reflect this purpose.

It is easy to derive the *m*GREG estimator of the total t_d under the *STSI* design by using the general expressions (6.2) and (6.3).

Element d' in $\mathbf{b}_d$ is estimated by

$$b_{dd'} = \frac{\hat{t}^*_{dd'}}{\hat{t}^*_{xd'}} = \frac{\sum_{h=1}^{H} \hat{t}^*_{dd'h}}{\sum_{h=1}^{H} \hat{t}^*_{xd'h}} \tag{6.4}$$

where $$\hat{t}^*_{dd'h} = \frac{N_h}{n_h} \sum_{s_h} z_{d'k} \zeta_{dk} y_k \tag{6.5}$$

and $$\hat{t}^*_{xd'h} = \frac{N_h}{n_h} \sum_{s_h} z_{d'k} x_k \tag{6.6}$$

Then the *m*GREG-estimator of t_d is

$$\hat{t}_{dr} = \sum_{d'=1}^{D} \hat{t}_{axd'} \frac{\hat{t}^*_{dd'}}{\hat{t}^*_{xd'}} \tag{6.7}$$

where

$$\hat{t}_{axd'} = \sum_{h=1}^{H} \frac{N_h}{n_{ah}} \Sigma_{s_{ah}} z_{d'k} x_k \tag{6.8}$$

The approximate variance of $\hat{t}_{dr}$ is

$$AV(\hat{t}_{dr}) = \sum_{h=1}^{H} \left\{ \frac{N_h^2}{n_{ah}} \left(1 - \frac{n_{ah}}{N_h}\right) S_{ydh}^2 + \frac{N_h^2}{n_h} \left(1 - \frac{n_h}{n_{ah}}\right) E_{STSI} E_{\mathbf{X}} \left[s_{Edh}^2 \right] \right\} \tag{6.9}$$

where $$s_{Edh}^2 = \frac{1}{n_{ah} - 1} \Sigma_{s_{ah}} \left(E_{dk} - \overline{E}_{dh} \right)^2 \tag{6.10}$$

$$E_{dk} = y_{dk} - \mathbf{x}'_k \, \mathbf{b}_{ad} \; ; \; \overline{E}_{dh} = \frac{1}{n_{ah}} \Sigma_{s_{ah}} E_{dk} \tag{6.11}$$

$$S_{ydh}^2 = \frac{1}{N_h - 1} \Sigma_{U_h} (y_{dk} - \tilde{y}_{dh})^2 \; ; \; \tilde{y}_{dh} = \frac{1}{N_h} \Sigma_{U_h} y_{dk} \tag{6.12}$$

and element d' in $\mathbf{b}_{ad}$ is calculated as

$$b_{add'} = \sum_{h=1}^{H} \frac{N_h}{n_{ah}} \Sigma_{s_{ah}} z_{d'k} y_{dk} / \hat{t}_{axd'} \tag{6.13}$$

A variance estimator is given by

$$\hat{V}(\hat{t}_{dr}) = \sum_{h=1}^{H} \left\{ \frac{N_h^2}{n_{ah}} \left(1 - \frac{n_{ah}}{N_h}\right) s_{ydh}^2 + \frac{N_h^2}{n_h} \left(1 - \frac{n_h}{n_{ah}}\right) s_{edh}^2 \right\} \tag{6.14}$$

where $$s_{ydh}^2 = \frac{1}{n_h - 1} \Sigma_{s_h} \left(y_{dk} - \hat{\tilde{y}}_{dh} \right)^2 ; \; \hat{\tilde{y}}_{dh} = \frac{1}{n_h} \Sigma_{s_h} y_{dk} \tag{6.15}$$

and $$s_{edh}^2 = \frac{1}{n_h - 1} \Sigma_{s_h} \left(e_{dk} - \bar{e}_{dh} \right)^2 \tag{6.16}$$

where $$e_{dk} = g_k (y_{dk} - b_{dd'} x_{d'k}), \, k \rightarrow d' \; ; \; \bar{e}_{dh} = \frac{1}{n_h} \Sigma_{s_h} e_{dk} \tag{6.17}$$

and $$b_{dd'} = \sum_{h=1}^{H} \frac{N_h}{n_h} \Sigma_{s_h} z_{d'k} y_{dk} / \hat{t}_{xd'}^{*} \tag{6.18}$$

The g_k may be set at 1 for every $k \in s$ or set at $\hat{t}_{axd'} / \hat{t}_{xd'}^{*}$ when $k \rightarrow d'$.

Remark 6.1. Consider the situation when the second-phase sample is taken according to the *SI* design without regard to the stratification in the first-phase design. This is not an unusual situation in practice. In many surveys a second-phase sample is not planned for in advance, but during the collection of data on the elements in the sample s_a one encounters measurement problems.

The measurement errors may then be studied by taking a second-phase sample and using a true device for the measurements and classification. Although the second-phase sample often primarily is taken to study the measurement process, not to arrive at better estimates, it can as well be used to get approximately unbiased estimates of the parameters of interest in the original survey to get some idea of the bias due to the measurement and classification errors in the estimates from the first-phase sample estimates.

The estimate $\hat{t}_{axd'}$ is then computed according to (6.8) while

$$\hat{t}^*_{dd'} = \frac{n_a}{n} \sum_{h=1}^{H} \frac{N_h}{n_{ah}} \Sigma_{s_h} z_{d'k} y_{dl} \quad ; \quad \hat{t}^*_{xd'} = \frac{n_a}{n} \sum_{h=1}^{H} \frac{N_h}{n_{ah}} \Sigma_{s_h} x_{d'k} \tag{6.19}$$

and the total t_d is estimated according to (6.7).

The approximate variance of $\hat{t}_{dr}$ is then given by

$$AV(\hat{t}_{dr}) = \sum_{h=1}^{H} \frac{N_h^2}{n_{ah}} \left(1 - \frac{n_{ah}}{N_h}\right) S_{ydh}^2 + E_{STSI} E_{\mathbf{X}} \left[\frac{n_a^2}{n} \left(1 - \frac{n}{n_a}\right) s_{Ed}^2 \right] \tag{6.20}$$

where

$$s_{Ed}^2 = \frac{1}{n_a - 1} \left\{ \sum_{h=1}^{H} \left(\frac{N_h}{n_{ah}}\right)^2 \Sigma_{s_{ah}} E_{dk}^2 - \frac{1}{n_a} \left(\sum_{h=1}^{H} \frac{N_h}{n_{ah}} \Sigma_{s_{ah}} E_{dk} \right)^2 \right\} \tag{6.21}$$

and E_{dk} is defined by (6.11).

A variance estimator is given by

$$\hat{V}(\hat{t}_{dr}) = \sum_{h=1}^{H} \frac{N_h^2}{n_{ah}} \left(1 - \frac{n_{ah}}{N_h}\right) s_{ydh}^2 + \frac{n_a^2}{n} \left(1 - \frac{n}{n_a}\right) s_{ed}^2 \tag{6.22}$$

where s_{ydh}^2 is defined by (6.15) and

$$s_{ed}^2 = \frac{1}{n-1} \left\{ \sum_{h=1}^{H} \left(\frac{N_h}{n_{ah}}\right)^2 \Sigma_{s_h} e_{dk}^2 - \frac{1}{n} \left(\sum_{h=1}^{H} \frac{N_h}{n_{ah}} \Sigma_{s_h} e_{dk} \right)^2 \right\} \tag{6.23}$$

where e_{dk} is defined by (6.17). □

6.3 Cluster Sampling and Two-Stage Sampling

Consider the following situation and notations. The population of elements $U=\{1, \ldots, k, \ldots, N\}$ is partitioned into N_{I} clusters $U_1, \ldots, U_i, \ldots, U_{N_{\mathrm{I}}}$, sometimes called primary sampling units (PSUs). The size of U_i is denoted N_i. We have

$$U = \bigcup_{i=1}^{N_{\mathrm{I}}} U_i \text{ and } N = \sum_{i=1}^{N_{\mathrm{I}}} N_i \tag{6.24}$$

The notation for the population of clusters is $U_{\mathrm{I}}=\{1, \ldots, i, \ldots, N_{\mathrm{I}}\}$. At stage one, a sample, $s_{a\mathrm{I}}$ of size $n_{a\mathrm{I}}$ of PSUs is selected from U_{I} according to the design $p_{a\mathrm{I}}(\cdot)$ with the first- and second-order inclusion probabilities $\pi_{a\mathrm{I}i}$, $\pi_{a\mathrm{I}ij}$.

When data are collected from every element in the selected clusters we talk about cluster sampling.

When a subsample is taken from the elements in the primary sample, *invariance* and *independence* (see Särndal *et al.* 1992) are required in the present section. Invariance means that the same design is used for the secondary sample of elements in a cluster each time that specific cluster is included in the primary sample, while independence means that the subsampling in one PSU is carried out independently of subsampling in any other PSU.

The sampling units at the second stage (SSUs) are the population elements. Given that U_i is chosen at stage one, a sample s_{ai} of size n_{ai} is selected from U_i according to the design $p_{ai}(\cdot)$ with inclusion probabilities $\pi_{ak|i}$ and $\pi_{akl|i}$. The sample of elements and its size are denoted

$$s_a = \bigcup_{i \in s_{a\mathrm{I}}} s_{ai} \text{ and } n_a = \sum_{i \in s_{a\mathrm{I}}} n_{ai} \tag{6.25}$$

Each selected element in the final sample s_a is classified into one of D classes and measured with respect to y by using the fallible procedure F. As before, a second-phase sample s of size n is taken from the first-phase sample s_a and the correct procedure T is applied to every element in the sample s.

I will only consider misclassification of elements although it is also possible to handle misclassification of clusters. Consider for example a household survey where households are the PSUs and household members are the SSUs. In this situation it may be of interest to classify the households (=clusters) as well as the members (=elements), and both classifications can be erroneous.

When the first-phase sample is taken under a cluster design or a two-stage design, the second-phase sample can be selected in one of several ways, for example:

i) Mimic the first-phase-design and take a subsample of the first-phase clusters and then subsample among the first-phase elements in the clusters selected in the second phase.

ii) Take the subsample of elements among the first-phase elements within *each* first-phase cluster, i.e. every cluster in the first-phase sample $s_{a\text{I}}$ has inclusion probability 1 in the second phase.

iii) Take the subsample of elements among the first-phase elements, independently of the first-phase design.

Formally for *i)* and *ii)*, in the second phase, a subsample of PSUs, s_{I} of size n_{I}, is selected from $s_{a\text{I}}$ according to the design $p_{\text{I}}(\cdot)$, with associated inclusion probabilities $\pi_{\text{I}i|a\text{I}}$ and $\pi_{\text{I}ij|a\text{I}}$. Given that PSU U_i is chosen in the second phase, a sample s_i of size n_i is selected from s_{ai} according to the design $p_i(\cdot \mid s_{ai})$ and with the inclusion probabilities $\pi_{k|ai}$ and $\pi_{kl|ai}$.

In *iii)*, the second-phase sample of the elements is taken directly from the first-phase sample of elements, s_a, disregarding the cluster level. There are pros and cons with the different ways of subsampling. For example, if the true procedure involves reinterviews it may be economical to mimic the first-phase design. On the other hand, if it is possible to identify certain groups of error-prone elements it might be more economical and maybe also more efficient to take a subsample of elements without regard to the cluster level.

In Särndal *et al.* (1992) expressions for the GREG estimator under cluster and two-stage designs are given when model ξ in (6.1) applies.

In this section the expressions for the *m*GREG estimator are given under the model (6.1) when the first-phase sample of elements is taken under a cluster or two-stage design and a second-phase sample of elements is taken according to *i)* or *ii)* above.

The *m*GREG estimator of the total t_d is then

$$\hat{t}_{dr} = \sum_{d'=1}^{D} \hat{t}_{axd'} \frac{\hat{t}^*_{ydd'}}{\hat{t}^*_{xd'}} = \sum_{d'=1}^{D} \hat{t}_{axd'} b_{dd'} \tag{6.26}$$

where

$$\hat{t}^*_{ydd'} = \sum_{s_\text{I}} \frac{\hat{t}^*_{ydd'i}}{\pi^*_{\text{I}i}} \quad ; \quad \hat{t}^*_{ydd'i} = \sum_{s_i} \frac{z_{d'k} y_{dk}}{\pi^*_{k|i}} \tag{6.27}$$

$$\hat{t}^*_{xd'} = \sum_{s_\text{I}} \frac{\hat{t}^*_{xd'i}}{\pi^*_{\text{I}i}} \quad ; \quad \hat{t}^*_{xd'i} = \sum_{s_i} \frac{x_{d'k}}{\pi^*_{k|i}} \tag{6.28}$$

and

$$\hat{t}_{axd'} = \sum_{s_{a\mathrm{I}}} \frac{\hat{t}_{axd'i}}{\pi_{a\mathrm{I}i}} \quad ; \quad \hat{t}_{axd'i} = \sum_{s_{ai}} \frac{x_{d'k}}{\pi_{ak|i}} \tag{6.29}$$

with $\pi^*_{\mathrm{I}i} = \pi_{a\mathrm{I}i}\pi_{\mathrm{I}i|a\mathrm{I}}$ and $\pi^*_{k|i} = \pi_{ak|i}\pi_{k|ai}$.

The expression for the approximate variance of an estimated total is decomposed into two components, $AV(\hat{t}_{dr}) = A + B$, with

$$A = \sum\sum_{U_\mathrm{I}} \left(\pi_{a\mathrm{I}ij} - \pi_{a\mathrm{I}i}\pi_{a\mathrm{I}j}\right) \frac{t_{di}}{\pi_{a\mathrm{I}i}} \frac{t_{dj}}{\pi_{a\mathrm{I}j}} + \sum_{U_\mathrm{I}} \frac{V_{di}}{\pi_{a\mathrm{I}i}} \tag{6.30}$$

where $$t_{di} = \sum_{U_i} y_{dk} \quad ; \quad V_{di} = \sum\sum_{U_i} \left(\pi_{akl|i} - \pi_{ak|i}\pi_{al|i}\right) \frac{y_{dk}}{\pi_{ak|i}} \frac{y_{dl}}{\pi_{al|i}} \tag{6.31}$$

and $$B = E_{p_a} E_{\mathbf{X}} \left[\sum\sum_{s_{a\mathrm{I}}} \left(\pi_{\mathrm{I}ij|a\mathrm{I}} - \pi_{\mathrm{I}i|a\mathrm{I}}\pi_{\mathrm{I}j|a\mathrm{I}}\right) \frac{\hat{t}_{Edi}}{\pi^*_{\mathrm{I}i}} \frac{\hat{t}_{Edj}}{\pi^*_{\mathrm{I}j}} + \sum_{s_{a\mathrm{I}}} \frac{V_{Edi}}{\pi_{a\mathrm{I}i}\pi^*_{\mathrm{I}i}} \right] \tag{6.32}$$

where $$\hat{t}_{Edi} = \sum_{s_{ai}} \frac{E_{dk}}{\pi_{ak|i}} \quad ; \quad V_{Edi} = \sum\sum_{s_{ai}} \left(\pi_{kl|ai} - \pi_{k|ai}\pi_{l|ai}\right) \frac{E_{dk}}{\pi^*_{k|i}} \frac{E_{dl}}{\pi^*_{l|i}} \tag{6.33}$$

and $$E_{dk} = y_{dk} - \mathbf{x}'_k \mathbf{b}_{ad} \tag{6.34}$$

An estimator of the variance is $\hat{V}(\hat{t}_{dr}) = \hat{A} + \hat{B}$, with

$$\begin{aligned} \hat{A} = {} & \sum\sum_{s_\mathrm{I}} \frac{\pi_{a\mathrm{I}ij} - \pi_{a\mathrm{I}i}\pi_{a\mathrm{I}j}}{\pi^*_{\mathrm{I}ij}} \frac{\hat{t}^*_{di}}{\pi_{a\mathrm{I}i}} \frac{\hat{t}^*_{dj}}{\pi_{a\mathrm{I}j}} + \\ & + \sum_{s_\mathrm{I}} \frac{\hat{V}_{di}}{\pi_{a\mathrm{I}i}\pi^*_{\mathrm{I}i}} - \sum_{s_\mathrm{I}} \frac{1}{\pi^*_{\mathrm{I}i}} \left(\frac{1}{\pi_{a\mathrm{I}i}} - 1 \right) \hat{V}^*_{di} \end{aligned} \tag{6.35}$$

where $\pi^*_{\mathrm{I}ij} = \pi_{a\mathrm{I}ij}\pi_{\mathrm{I}ij|a\mathrm{I}}$ and

$$\hat{t}^*_{di} = \sum_{s_i} \frac{y_{dk}}{\pi^*_{k|i}} \quad ; \quad \hat{V}_{di} = \sum\sum_{s_i} \frac{\pi_{akl|i} - \pi_{ak|i}\pi_{al|i}}{\pi^*_{kl|i}} \frac{y_{dk}}{\pi_{ak|i}} \frac{y_{dl}}{\pi_{al|i}} \tag{6.36}$$

with $\pi^*_{kl|i} = \pi_{akl|i}\pi_{kl|ai}$ and

$$\hat{V}^*_{di} = \sum\sum_{s_i} \frac{\pi^*_{kl|i} - \pi^*_{k|i}\pi^*_{l|i}}{\pi^*_{kl|i}} \frac{y_{dk}}{\pi^*_{k|i}} \frac{y_{dl}}{\pi^*_{l|i}} \tag{6.37}$$

An estimator of component B is

$$\hat{B} = \sum\sum_{s_{\mathrm{I}}} \frac{\pi_{\mathrm{I}ij|a\mathrm{I}} - \pi_{\mathrm{I}i|a\mathrm{I}}\pi_{\mathrm{I}j|a\mathrm{I}}}{\pi_{\mathrm{I}ij|a\mathrm{I}}} \frac{\hat{t}^*_{edi}}{\pi^*_{\mathrm{I}i}} \frac{\hat{t}^*_{edj}}{\pi^*_{\mathrm{I}j}} + \sum_{s_{\mathrm{I}}} \frac{\hat{V}_{edi}}{\pi_{a\mathrm{I}i}\pi^*_{\mathrm{I}i}} \tag{6.38}$$

where $\hat{t}^*_{edi} = \sum_{s_i} \frac{g_k e_{dk}}{\pi^*_{k|i}}$; $\hat{V}_{edi} = \sum\sum_{s_i} \frac{\pi_{kl|ai} - \pi_{k|ai}\pi_{l|ai}}{\pi_{kl|ai}} \frac{g_k e_{dk}}{\pi^*_{k|i}} \frac{g_l e_{dl}}{\pi^*_{l|i}}$ (6.39)

with $e_{dk}=y_{dk}-\mathbf{x}'_k\mathbf{b}_d$ and g_k is 1 for every k or equal to $\hat{t}_{axd'}/\hat{t}^*_{xd'}$ when $k \rightarrow d'$.

Two-phase sampling in combination with cluster or two-stage sampling and the use of auxiliary information does lead to rather cumbersome formulas.

CHAPTER 7

An Application

7.1 Background

In this chapter a survey taken at Statistics Sweden will be used as an example of the two-phase sampling technique when the data are impaired by measurement errors and misclassification. The survey was taken as a part of the planning of the 1990 Population and Housing Census in Sweden (FoB90). In FoB90, data on individuals, households and dwellings are of interest. In the example presented here only data on households will be used.

7.2 Description of the Pilot Study

In the autumn of 1989 a pilot study, named PuF, was carried out to test the questionnaire to be used in FoB90.

The population of interest was the households in Sweden as of the November 1 1989 that contained at least one member in the age-range 16-64 years.

A household in FoB90 is defined as a person or a group of persons registered in the same municipality and living in the same dwelling. Persons that temporarily live outside the municipality, e.g. due to studies, military service, being in hospital, etc., are also considered as members of the household.

Parameters of interest in this example are the number of households with 1, 2, 3, 4, or 5 or more (5+) members.

Let U_{I} be the population of households that is of interest and let ζ_{di} be defined as in chapter 1, i.e. $\zeta_{di}=1$ if household i is of size d (d=1, ..., 5), and $\zeta_{di}=0$ otherwise. Then the total number of households in domain d is $t_d = \sum_{U_{\mathrm{I}}} \zeta_{di}$.

In Sweden, no reliable sampling frame of households exists. However, the register of the total population, RTB, is a very reliable register that contains practically every person living in Sweden. It was decided to use RTB as sampling frame for the PuF sample by taking a sample of individuals in the age-range 16-64 years.

7.2.1 Estimating Household Parameters from a Sample of Individuals

Since the sample taken was a sample of individuals it cannot be used to estimate the number of households without considering the number of members of the household to which the selected individual belongs. This follows from the fact that a household with many members in the age-range 16-64 years has a larger probability of being represented in the sample than a household with fewer members in the same age-range. The problem can be solved in different ways. For example,

1) Compute the first- and second-order inclusion probability of the household *ex post* with regard to the number of members, i.e. the sample is considered to be a πps-sample of households. This can be done when data about the number of household members in the 16-64 age-range are collected from the households that are obtained *through* the sample of individuals.

   Let s_{I} be the sample of different households obtained through the sample of individuals and let $\pi_{\mathrm{I}i}$ be the *ex post* inclusion probability of household i. An unbiased estimator of t_d is then $\hat{t}_{\mathrm{I}d} = \sum_{s_{\mathrm{I}}} \zeta_{di} / \pi_{\mathrm{I}i}$.

2) Use an appropriate weight for each individual in the sample. This approach was suggested by Hansen, Hurwitz & Madow (1953). The problem of finding optimal weights when the *STSI* design is used was studied by Rosén (1987).

   Let s be the sample of individuals, let individual k belong to a household with a_k members in the age-range 16-64 years and let π_k be the inclusion probability of individual k, then *one* proper weight is $w_k = 1/a_k$ and an unbiased estimator for t_d is $\hat{t}_{wd} = \sum_s w_k \zeta_{dk} / \pi_k$.

Both estimators $\hat{t}_{\mathrm{I}d}$ and $\hat{t}_{wd}$ need reliable data on the number of household members in the age-range 16-64 years, but $\hat{t}_{wd}$ is easier to use in practice when the sample of individuals is taken by a standard design, e.g. the *SI* or *STSI* design, and it was the one preferred in PuF.

7.2.2 Sample Design, Data Collection and Estimation in PuF

The sampling frame was divided into two strata. Stratum 1 consisted of individuals living in the three largest cities in Sweden, Stockholm, Gothenburg and Malmö, and stratum 2 consisted of individuals living outside these areas. The sample sizes were 1,000 in stratum 1 and 2,000 in stratum 2.

A questionnaire was sent by mail to the individuals in the sample, who were asked to answer the questions and return the questionnaire to Statistics Sweden. A total of 628 questionnaires were returned in stratum 1 and 1,376 were returned in stratum 2, i.e. the response rates were 62.8% and 68.8% respectively.

The respondents were considered to be a simple random sample from the initial samples in each stratum although there was no evidence for that. This means that the respondents will be treated as an *STSI* sample $s_a=\{s_{a1}, s_{a2}\}$ with sample sizes n_{a1}=628 and n_{a2}=1,376. If the respondents are not simple random samples from the initial samples there is a risk that the estimates will be biased due to the nonresponse.

If the respondents do not answer the questionnaire correctly as far as the number of household members and their ages are concerned, two type of errors will occur:

1) Classification errors, since the household will be allocated to the wrong domain due to too many or too few members in the household according to the questionnaire.
2) Measurement errors, since the weights will not be correct when the number of households in domain d is estimated.

For respondent k in the sample s_a two values were obtained: 1) a possibly erroneous value on vector $\mathbf{z}_k$ of length 5 with one non-zero element z_{dk}, equal to "1" if the household contains d members according to the questionnaire, and 2) a possibly erroneous value $x_k=1/b_k$, where b_k is the number of members of the household in the 16-64 age-range according to the questionnaire.

The number of households in domain d is estimated by

$$\hat{t}_{axd} = \sum_{h=1}^{2} \frac{N_h}{n_{ah}} \sum\nolimits_{s_{ah}} z_{dk} x_k \tag{7.1}$$

where N_h is the size of stratum h.

It is likely to be the case that the number of small households will be overestimated and the number of large households will be underestimated, and hence the total number of households will be overestimated when (7.1) is calculated with data from s_a. This is because the respondents often fail to include those members that are temporarily absent. Another reason for erroneous answers is that people living together without being married do not always consider themselves as *one* household despite the fact that the definition used in FoB90 says that they should.

7.2.3 A Second-Phase Sample

To evaluate the collected values from the pilot study, a subsample was selected within each stratum using the *SI* design. The subsample sizes were 400 and 600 in stratum 1 and 2, respectively, and the subsamples were selected before the questionnaires were sent out. 624 respondents in the first-phase sample turned out to belong to the evaluation sample s. The evaluation survey was called EPuF.

The numbers of respondents in the evaluation sample s were n_1=206 and n_2=336 in strata 1 and 2, respectively. The respondents in s were interviewed on the telephone, and a more comprehensive questionnaire than the original one was used. The answers in the telephone interview were judged to be very reliable and are here considered to be true. In the same way as in the first-phase data collection, the nonresponse in phase two was assumed to occur at random within each stratum.

Now the following situation exists. The collected data in the first phase may contain misclassification and/or measurement errors due to an error-prone procedure F, while in the second phase the error-free procedure T gives the correct number of household members and the correct number of members in the 16-64 age-range. For each respondent in the second-phase sample, the values of the vector $\mathbf{z}_k$ and the variable x_k are known from PuF and the values of the vector ζ_k and the variable $y_k=1/a_k$ are known from EPuF, where a_k is the true number of members in the 16-64 age-range in the household that contains individual k.

The number of households in domain d may be estimated by use of the π* estimator for t_d based on the data from sample s in the second-phase sample only. The number of households that belong to domain d is then estimated by

$$\hat{t}_d^* = \sum_{h=1}^{2} \frac{N_h}{n_h} \sum_{s_h} \zeta_{dk} y_k \tag{7.2}$$

where s_h is the set of respondents in stratum h.

7.2.4 Combining the First- and Second-Phase Samples — The mGREG Estimator

When the data from the 2,004 respondents in the first phase are used as auxiliary information, we hope that estimates with smaller standard errors will be obtained than those based on (7.2). The number of households in domain d is estimated by the mGREG estimator,

$$\hat{t}_{dr} = \sum_{d'=1}^{5} \hat{t}_{axd'} \frac{\hat{t}_{dd'}^*}{\hat{t}_{xd'}^*} \tag{7.3}$$

where $\hat{t}_{axd'}$ is defined by (7.1),

$$\hat{t}_{dd'}^* = \sum_{h=1}^{2} \frac{N_h}{n_h} \sum_{s_h} z_{d'k} \zeta_{dk} y_k \text{ and } \hat{t}_{xd'}^* = \sum_{h=1}^{2} \frac{N_h}{n_h} \sum_{s_h} z_{d'k} x_k \tag{7.4}$$

7.3 Results

The results of the calculations in sections 7.2.2-7.2.4 are given in table 7.1. The figures in italics are the estimates of the standard errors computed as the square root of the variance $\hat{V}(\hat{t})$ for each estimator.

The results of the calculations are compared with the results of the evaluation study of FoB90, EU90 (Statistics Sweden 1992), which was carried out one year later.

It is easily seen from table 7.1 that the unadjusted estimator heavily underestimates the number of households with more than 3 members and overestimates the number of small households compared to the estimates obtained with π^* and *m*GREG.

The π^* estimates are rather close to EU90, while *m*GREG seems to overestimate the number of small households and thus to overestimate the total number of households. The reason for this is probably that the large nonresponse in PuF is not random. People living alone or in two-person households are more likely not to respond to the questionnaire. It may be noted that there was nonresponse in EU90 too, the nonresponse rate being about 16%.

Generally, the use of auxiliary variables in the *m*GREG estimator will reduce selection bias when the assumed response model (random nonresponse within each stratum) is erroneous if they are strong predictors of the response variable of interest (Särndal & Swensson 1987, Bethlehem 1988). In the application described here this means that the selection bias in the second phase should be small while it may remain large in the first phase. When the π^* estimator is used the possible biases in the first and second phases may partly eliminate each other.

The estimated standard errors are smallest when the unadjusted estimator is used, as expected, while the estimated standard errors from *m*GREG are generally smaller than those of the π^* estimator, especially for the estimate of the number of households with one member. For the estimates of the number of households with 3 or more members, the benefit is moderate or even absent, probably due to low correlation between x_{dk} and y_{dk}.

7.4 Estimating the Simple Response Variance

When the erroneous values in the first-phase sample s_a are used to calculate the variances of the unadjusted estimator, one component is the simple response variance described in chapter 1. This component is due to the measurement errors and misclassification and can be computed from the second-phase sample according to the expression (2.8) in chapter 2,

$$\hat{G}_d = \sum_{h=1}^{2} \frac{N_h^2}{n_{ah} n_h} \sum_{s_h} \left(x_{dk} - y_{dk}\right)^2 \tag{7.5}$$

Then we get the following estimates of the relative sizes of $\hat{G}_d$, $\hat{RG}_d = \hat{G}_d / \hat{V}(\hat{t}_{axd})$:

d	1	2	3	4	5+	*Tot*
$\hat{RG}_d$	0.45	0.91	1.10	1.57	1.89	0.83

The estimates of the relative, simple response variances are large, even too large when $d \geq 3$, since they should be less then 1. The estimator (7.5) is not an unbiased estimator, and the bias may be large, especially in the domains, as can be seen from expression (2.10) in chapter 2. The estimates of the simple response variance are not considered to be reliable in this case.

Table 7.1 *A comparison between the estimates from the unadjusted estimator (PuF), the π* estimator, the mGREG estimator and EU90.*

	Size of Household					
Estimator	1	2	3	4	5+	*Total*
PuF	1,659,500	2,312,140	500,099	174,655	30,707	4,677,102
(7.1)	*65,479*	*52,477*	*30,402*	*19,003*	*8,709*	*40,874*
π*	1,315,217	1,319,200	425,226	540,042	188,518	3,788,203
(7.2)	*115,398*	*74,341*	*43,558*	*50,101*	*29,802*	*69,717*
*m*GREG	1,592,821	1,326,669	405,942	530,872	187,742	4,044,046
(7.3)	*76,380*	*68,286*	*42,319*	*49,255*	*30,008*	*56,146*
EU90	1,242,258	1,164,897	490,783	463,668	203,309	3,564,915
	25,324	*13,996*	*9,988*	*6,606*	*4,079*	*14,512*

CHAPTER 8

Allocation

8.1 Introduction and Preliminaries

In practice the available resources for a survey are always limited. The cost of the survey must not exceed a certain amount, there may be time limits, etc. The limited sampling resources are to be allocated as efficiently as possible at the planning stage of a survey.

When the situation described in chapter 1 exists, it is not self-evident that, for example, a two-phase sampling design makes the best use of the available resources.

Deming (1977) argues that when the budget is restricted, it may be better to use all resources on the true device T instead of using a two-phase design, unless 1) the cost per unit in the first phase is widely different from that in the second phase, and 2) the reliability of the false device F is high. The closer the per unit costs are in the first and second phases, the higher does the reliability need to be to justify the two-phase design. However, Deming (1977) also admits that there may be advantages with a large first-phase sample: it may for example bring out problems that no one could have foreseen.

Another reason for a two-phase design is that in most surveys there are several variables of interest and not all of them need to be error-prone. It may be of interest to study only the error-prone variables in the survey by taking a subsample from the first-phase sample and using a true device to get correct values for them.

When a two-phase sample design is used, the auxiliary information in the first-phase sample s_a may be used in either of at least two ways: (*i*) take a second-phase sample with the *SI* design and use the auxiliary information from the first phase in the *m*GREG estimator, or (*ii*) use the information in s_a to make an efficient stratification of the first-phase sample and use the *STSI* design in the second phase with the *m*GREG estimator.

In this chapter the focus will be on the determination of sample sizes in the following situation:

1) A first-phase sample s_a of size n_a is taken under an *SI* design. Let c_1 be the per unit sample cost in the first phase.

2) A second-phase sample is taken in either of two ways: (*i*) a second-phase *SI* sample, s, of size $n=\nu n_a$, is drawn from s_a, where $0<\nu\leq 1$ is an *a priori* fixed constant, or (*ii*) in the sample s_a, D classes are identified by the fallible device F, and these classes constitute the strata, i.e. $s_a=s_{a1}\cup\ldots s_{ad'}\cup\ldots\cup s_{aD}$, where $s_{ad'}$ ($d'=1, \ldots, D$) is of size $n_{ad'}$. The expectation of $n_{ad'}$ under model m in section 1.5 is $E_{\mathbf{X}}[n_{ad'}]=n_a\pi_{d'}$, where $\pi_{d'} = \sum_U \theta_{d'k}/N$.
   An *SI* subsample $s_{d'}$ of size $n_d=\nu_{d'}n_{ad'}$ is taken in stratum d', where $0<\nu_{d'}\leq 1$ are *a priori* fixed constants. The true device T is applied to the units in the second phase sample.
3) The *m*GREG estimator is used to estimate the total t_d.

Simple random sampling in the second phase means that the sampling rate will be about the same in each identified class in the first-phase sample s_a. This may not be an efficient plan if the cost of true classification and/or correct measurements differs between the classes identified by the fallible device F in s_a, or if the amount of misclassification and/or measurement errors differs from class to class.

When a second-phase *STSI* sample is taken, a distinction has to be made between the situation when the stratification corresponds to a fixed division of the population U, as described in section 6.1, and the situation when the division into strata is done for the given first-phase sample s_a. The latter case, which will be treated in this chapter, is called "conditional resampling" (CR) by Haitovsky & Rapp (1992) when the classes identified in s_a are used as strata, since a unit k is allocated to stratum d' conditional on the observed, error-prone classification. In fact this is a special case of a two-phase sampling design.

Tenenbein (1970, 1971, 1972) treated the allocation problem when the *SI* design is used in both phases, $x_{dk}=z_{dk}$ and $y_{dk}=\zeta_{dk}$, i.e. when $\hat{t}_{dr}$ is an estimator of the number of units in domain d. Deming (1977), Smith (1989) and Haitovsky & Rapp (1992) treated the allocation problem in the CR case with the *SI* design in the first phase and the *STSI* design in the second phase when $\hat{t}_{dr}$ is an estimator of the number of units in domain d.

Deming (1977) claimed that optimum allocation in the second phase should not be used unless much is known about the measurement procedure; instead he favours proportional allocation, which he regards as simple to apply and foolproof.

When proportional allocation is used, the subsampling fraction, $\nu=n/n_a$, is the same in each stratum, where n is the total size of the second-phase sample. However, when the *m*GREG estimator is used it turns out that proportional allocation will give the same variance as when the *SI* design is used in the second phase.

Let $c_{2d'}$ be the per unit second phase sample cost in stratum d'. The expected cost of sampling is assumed to be $EC = n_a c_1 + n_a \sum_{d'=1}^{D} \pi_{d'} c_{2d'} \nu_{d'}$ when the two-phase design *SI, STSI* is used. The cost function is simplified to $EC = n_a(c_1 + \nu c_2)$ when the second-phase design is *SI*, where c_2 is the per unit sample cost in the second phase.

It is assumed that n_a is so large that $n_{ad'}$=0 with probability zero and that $n_{d'}$ (or n) is at least equal to 1.

Recall from chapter 3 that when the two-phase design *SI, SI* is used, the approximate variance of the *m*GREG estimator of $\hat{t}_{dr}$ is

$$AV(\hat{t}_{dr}) = \frac{N^2}{n_a}\left(1 - \frac{n_a}{N}\right) S_{yd}^2 + E_{SI} E_{\mathbf{X}}\left[\frac{N^2}{n}\left(1 - \frac{n}{n_a}\right) s_{Ed}^2\right] = V_1 + V_2 \tag{8.1}$$

If ν is a fixed constant, the component V_2 may be written

$$V_2 = N^2 \frac{1}{n_a}\left(\frac{1}{\nu} - 1\right) E_{SI} E_{\mathbf{X}}\left[s_{Ed}^2\right] \tag{8.2}$$

However, (8.2) is not a suitable expression when the allocation problem is of interest. The component V_2 has to be cast into an explicit form that allows the determination of n_a and ν.

The approximation $E_{SI} E_{\mathbf{X}}\left[s_{Ed}^2\right] \approx S_{yd}^2\left(1 - R_d^2\right)$ is used in (8.2), where R_d^2 is recognised as the coefficient of reliability in section 3.3. Then (8.1) is approximated by

$$AV(\hat{t}_{dr}) \approx \frac{N^2}{n_a}\left(1 - \frac{n_a}{N}\right) S_{yd}^2 + \frac{N^2}{n_a}\left(\frac{1}{\nu} - 1\right) S_{yd}^2\left(1 - R_d^2\right) = V_1 + AV_2 \tag{8.3}$$

which is a more suitable expression when the allocation problem is considered.

When the two-phase design is *SI, STSI* and $\nu_{d'}$ (d'=1, …, D) are fixed constants, the approximate variance of the *m*GREG estimator of $\hat{t}_{dr}$ is

$$AV(\hat{t}_{dr}) = V_1 + \frac{N^2}{n_a} \sum_{d'=1}^{D}\left(\frac{1}{\nu_{d'}} - 1\right) E_{SI} E_{\mathbf{X}}\left[\frac{n_{ad'}}{n_a} s_{Edd'}^2\right] = V_1 + V_2 \tag{8.4}$$

The expected value $E_{SI} E_{\mathbf{X}}[\cdot]$ in (8.4) is approximated by, (see Appendix B),

$$E_{SI} E_{\mathbf{X}}\left[\frac{n_{ad'}}{n_a} s_{Edd'}^2\right] \approx \pi_{d'} S_{Edd'}^2 = S_{ydd'}^2 + b_{Udd'}^2 S_{xd'}^2 - 2 b_{Udd'} S_{ydxd'} \tag{8.5}$$

where the expressions for $S_{ydd'}^2$, $S_{xd'}^2$, $S_{ydxd'}$ and $b_{Udd'}$ are given in Appendix B. Then V_2 in (8.4) is approximated by

$$AV_2 = \frac{N^2}{n_a} \sum_{d'=1}^{D} \pi_{d'} \left(\frac{1}{\nu_{d'}} - 1 \right) S^2_{Edd'} \tag{8.6}$$

Remark 8.1. Let the two-phase design be *SI*, *STSI* with the D classes identified by device F in s_a, as strata, let the measurement model in section 1.5 be applicable with $y_{dk}=\zeta_{dk}$, $x_{dk}=z_{dk}$, $\mu^{(1)}_{xk}=1$, $\sigma^2_{xk}=0$, $\theta_{d'k}=\theta_{d'|d}$, $k\in U_d$, and let $P_d = \sum_U \zeta_{dk}/N$. Then (8.5) is approximately equal to $P_d\theta_{d'|d} - (P_d\theta_{d'|d})^2/\pi_{d'}$.

The approximate variance of the *m*GREG-estimator

$$\hat{t}_{dr} = \frac{N}{n_a} \sum_{d'=1}^{D} \frac{\hat{t}_{ydd'}}{\hat{t}_{xd'}} \hat{t}_{axd'} \tag{8.7}$$

may then be shown to be

$$AV(\hat{t}_{dr}) \approx \frac{N^2 P_d}{n_a} \left((1-P_d)R_d^2 + \sum_{d'=1}^{D} \frac{\theta_{d'|d}}{\nu_{d'}} - P_d \sum_{d'=1}^{D} \frac{\theta^2_{d'|d}}{\pi_{d'}\nu_{d'}} \right) - NP_d(1-P_d) \tag{8.8}$$

where R_d^2 is recognised as the coefficient of reliability in section 3.3.

The expression (8.8) is the same as expression (1.7) in Haitovsky & Rapp (1992) apart from the term $NP_d(1\text{-}P_d)$ which comes from the fpc in V_1. □

Now, the problem is to find n_a and $\nu_{d'}$ (or ν) such that 1) the variance of $\hat{t}_{dr}$ is minimised at a fixed total cost C_0, or 2) the total expected cost EC is minimised at a fixed variance V_* of $\hat{t}_{dr}$.

8.2 Minimising the Variance at a Fixed Cost

The allocation problem is called multicharacter when there are several totals t_d (d=1, ..., D) to be estimated. Minimising $AV(\hat{t}_{dr})$ subject to the constraint $EC=C_0$ will usually give different values, $n_a(d)$ and $\nu_{d'}(d)$ (or $\nu(d)$), for different d-values. Then for which d-value shall $n_a(d)$ and $\nu_{d'}(d)$ (or $\nu(d)$) be chosen?

There does not seem to be a natural solution to this problem. When there is *one* total t_d that is deemed more important than the others, that total may be used in the minimisation. Another approach, suggested by Smith (1989), is to minimise the variance function $V=\sum_{d=1}^{D} a_d AV(\hat{t}_{dr})$ subject to the cost constraint $EC=C_0$. The weights a_d may then be chosen to reflect the importance of t_d. This removes the multicharacter complication from the problem and the solution can be obtained by minimising $EC\times\sum_{d=1}^{D} a_d AV(\hat{t}_{dr})$, subject to the constraint $EC=C_0$.

The optimum subsampling fraction in stratum d', $\nu_{d'}(d)$, (d'=1, ..., D) for a given d-value turns out to be

$$\nu_{d'}(d)=\sqrt{c_1 S^2_{Edd'} \Big/ c_{2d'}\left(S^2_{yd}-\sum_{d'=1}^{D}\pi_{d'}S^2_{Edd'}\right)} \qquad \text{if } 0<\nu_{d'}(d)\le 1 \qquad (8.9)$$

and the optimal $n_a(d)$ is

$$n_a^0(d)=\frac{C_0}{c_1}\left\{1+\sum_{d'=1}^{D}\frac{c_{2d'}}{c_1}\nu_{d'}(d)\pi_{d'}\right\}^{-1} \qquad \text{if } 1\le n_a^0(d)\le N \qquad (8.10)$$

When the second-phase design is *SI*, (8.9) and (8.10) are simplified to

$$\nu(d)=\sqrt{\frac{\left(1-R_d^2\right)}{R_d^2}\frac{c_1}{c_2}} \qquad \text{if } 0<\nu(d)\le 1 \qquad (8.11)$$

and

$$n_a^0(d)=\frac{C_0}{c_1+c_2\nu(d)} \qquad \text{if } 1\le n_a^0(d)\le N \qquad (8.12)$$

Certainly, a lot of *a priori* information about the survey and the measurement process is needed to compute the optimum allocation solution. Tenenbein (1971) discusses the allocation problem when the two-phase sample design is *SI, SI*. When nothing or very little is known about R_d^2, he suggests a three-phase sampling procedure, where the first-phase sample is used to get estimates of R_d^2; however, the sample size must be rather large to get reliable estimates, as Deming (1977) points out. As an alternative to three-phase sampling, Tenenbein (1971) suggests that R_d^2=0.5 be used in (8.11).

Tenenbein (1971) has also studied the sensitivity of the optimum solution when an erroneous R_d^2 is used in the allocation and $\hat{t}_{dr}$ is an estimator of the number of units in domain d. He found that the approximate variance (8.3) is rather robust against errors in R_d^2, even when the error is as large as ±0.2.

8.3 Minimising the Expected Cost at a Fixed Variance

Problem 2) in section 8.1 can be formulated in the following way: minimise $EC = n_a c_1 + n_a \sum_{d'=1}^{D} \pi_{d'} c_{2d'} \nu_{d'}$ subject to the constraints $AV(\hat{t}_{dr}) \leq V_*(d)$.

When there is only *one* total, t_d, of interest, the optimal subsampling fraction $\nu_{d'}(d)$ is given by expression (8.9), while the optimal value of $n_a(d)$ is

$$n_a^*(d) = \frac{N^2}{\left(V_*(d) + NS_{yd}^2\right)} \left\{ S_{yd}^2 + \sum_{d'=1}^{D} \pi_{d'} \left(\frac{1}{\nu_{d'}(d)} - 1 \right) S_{Edd'}^2 \right\} \quad \text{if } 1 \leq n_a^*(d) \leq N \qquad (8.13)$$

When the second-phase design is *SI*, (8.13) is simplified to

$$n_a^*(d) = \frac{N^2 S_{yd}^2}{\nu(d)\left(V_*(d) + NS_{yd}^2\right)} \left(1 - R_d^2 + R_d^2 \nu(d)\right) \qquad \text{if } 1 \leq n_a^*(d) \leq N \qquad (8.14)$$

Tenenbein (1972) suggested that when several totals are of interest, the *d*-value with the smallest R_d^2-value should be chosen in (8.11) and (8.14). However, he found that the expected cost is sensitive to errors in R_d^2, especially when $R_d^2 > 0.7$.

Minimising the expected cost, *EC*, subject to $AV(\hat{t}_{dr}) \leq V_*(d)$ (d=1, …, D) offers a general solution. It can be described as the minimisation of a convex object function subject to a number of linear constraints.

Chromy (1987) suggested a general algorithm that minimises the total sampling cost by convex programming when there are several different variance constraints involved. The algorithm can be used for a two-phase sample scheme and allows restrictions on the sample fractions, like $0 < \nu_{d'} \leq 1$; it is also easy to implement the additional restriction $n_a \leq N$ in the algorithm.

Bethel (1989) suggested an algorithm that solves the problem of optimum allocation for a multicharacter survey. However, the algorithm is not suitable for two-phase designs.

Other algorithms are suggested by Armstrong & Wu (1992). They suggested a method that is tailor-made for two-phase sampling, but problems with the convergence have been reported. Armstrong & Wu (1992) also suggested a non-iterative method that gives an approximate solution to the optimisation problem.

When the expected cost is minimised under several variance constraints it might well happen that the obtained minimum cost does not exhaust the available budget, or, perhaps worse, the expected cost exceeds the budget. What is then to be done?

The obvious solution is to change the variance constraints. The optimal solution under new constraints can be obtained in the following way. Let the minimum expected cost C^* under the constraints $AV(\hat{t}_{dr}) \le V_*(d)$, ($d$=1, ..., D) be obtained for the values (n_a^*, ν), where n_a^* is the first-phase sample size and ν is a vector of D subsampling fractions with element d' equal to $\nu_{d'}$. Then $n_a^+ = n_a^* \times q$ is the optimal solution under the constraints $AV(\hat{t}_{dr}) \le V_*^+(d) = V_*(d)/q$, ($d$=1, ..., D), where $q = C_0/C^*$ is a constant such that the minimum cost is equal to the available budget C_0 with the solution (n_a^+, ν).

8.4 An Illustration of the Allocation Methods

In this section the performance of the different allocation methods described in sections 8.2 and 8.3 will be illustrated. The data are assumed to have the same characteristics as described in chapter 5 but only variables y_1 and y_2 are used. The relation between the sample costs in the first and second phase, c_1:c_2, is assumed to be 1:20 and the available budget is C_0=2,500.

The allocation of units when the two-phase sample design and the *m*GREG estimator are used is compared to the situation when only the true device T is used for the classifications and measurements and the π^* estimator is used for estimating t_d. The budget then allows for 125 sample units and, ignoring the fpc from the first-phase sample, the estimated total $\hat{t}_d$ will have the variances $AV(\hat{t}_1)$=26,880 in domain 1 and $AV(\hat{t}_2)$=43,520 in domain 2 for variables y_1 and y_2.

8.4.1 Using *SI* Designs in both Phases

Consider the case when the two-phase design *SI*, *SI* is used. From chapter 4 we have two misclassification cases, A) and B), two variables, y_1 and y_2, and two domains, 1 and 2. For each combination of misclassification case, variable and domain, the optimum allocation solution $(n_a^0(d), \nu(d))$ is computed, 1) by minimising $AV(\cdot)$ at the fixed cost C_0=2,500 giving the minimum variance $AV^0(d)$, and 2) by minimising the expected cost, *EC*, under the constraints $V_*(1)$=26,880 and $V_*(2)$=43,520, giving the optimum solutions $(n_a^*(d), \nu(d))$, and the minimum costs $C^*(d)$, d=1, 2. The solutions are given in table 8.1.

In practice, of course, only *one* of the four solutions in each case A) and B) is possible and therefore it is not possible to realise the gains simultaneously. Obviously there is a substantial gain in cost or variance for all variables and domains, compared to using the true device T only. When the two-phase scheme is used in combination with the *m*GREG-estimator, the gains are between 19% and 49%.

8.4.2 Using the *STSI* Design in the Second Phase

When the classes identified in the first-phase sample are used as strata in the second phase, the optimum solutions are calculated under the same cost and variance constraints as in section 8.4.1. The results are given in table 8.2 for each combination of misclassification case, variable and domain.

A comparison between the results in tables 8.1 and 8.2 shows that using the *STSI* design instead of the *SI* design in the second phase decreases the costs or variances by a further 0-15%-units as compared with $V_*(d)$ and C_0. Still there is the problem of selecting *one* of the triples ($n_a^0(d)$ (or $n_a^*(d)$), $\nu_1(d)$, $\nu_2(d)$). One possibility is to use the convex programming method when the fixed variance problem is of interest and minimises a variance function, e.g. $V = \sum_{d=1}^{D} AV(\hat{t}_{dr})$ when the fixed cost problem is of interest.

Solving the Fixed Cost Problem by a Variance Function

The problem is to calculate n_a^0 and $\nu_{d'}$ such that the variance function $V= \sum_{d=1}^{D} AV(\hat{t}_{dr})$ is minimised subject to the constraint $C=C_0$, where C_0 is the fixed budget. The problem is easily solved by using $\sum_{d=1}^{D} S_{yd}^2$ and $\sum_{d=1}^{D} S_{Edd'}^2$ instead of S_{yd}^2 and $S_{Edd'}^2$ in (8.9). The solutions n_a^0 and $\nu_{d'}$ for each combination of misclassification case and variable are given in table 8.3 and the calculated variances $AV^0(d)$ are given in table 8.4.

Solving the Fixed Variance Problem by Convex Programming

The expected cost was minimised for each combination of misclassification case and variable subject to the constraints

$$AV(\hat{t}_{dr}) \leq V_*(d) \qquad (8.15)$$

and

$$1 \leq n^*_{d'} \leq \pi_{d'} n^*_a \quad ; \quad 1 \leq n^*_a \leq N \qquad (8.16)$$

by using the algorithm suggested by Chromy (1987).

The obtained minimum cost C^* is smaller than C_0=2,500 in each combination of misclassification cases and variables, and the restriction (8.15) was recalculated to exhaust the available budget, whereby $q=C_0/C^*$, $V^+_*(d) = V_*(d) / q$. The recalculated sample sizes are $n^+_a = n^*_a \times q$ where n^+_a is the sample size that gives the minimum cost C_0 under the restrictions $AV(\hat{t}_{dr}) \leq V^+_*(d)$ (d=1, ..., D). The obtained variance $AV^+(d)$ at the minimum cost C_0 is calculated as $AV^+(d)=AV^*(d)/q$, where $AV^*(d)$ is the obtained variance at the minimum cost C^*. The optimal values (n^*_a, ν) and n^+_a are given in table 8.3 and the calculated and recalculated variances are given in table 8.4.

The expected costs are reduced by between 21% and 40% as compared to the solutions in table 8.1. The reduction is larger for variable y_2 than for y_1. Three of the four solutions are identical with the "domain solution" for the corresponding combination of misclassification and variable in table 8.2 with the largest cost.

The reductions in the variances compared with $V_*(d)$ are shown to be between 23% and 54% when C^*= C_0=2,500. It is remarkable how small the losses are when the variances are compared with the smallest possible values, given in table 8.2.

8.5 Comments on the Illustrations

The results of the illustrations can be summarised as follows:

i) Using the two-phase sampling scheme *SI*, *SI* in combination with the *m*GREG estimator, leads to a substantial reduction of the variances, when the budget is restricted, as compared to using only the true device T.

ii) Using the two-phase sampling scheme *SI*, *STSI* gives a further reduction of the variances for most combinations of misclassification, variables and domains.

iii) When convex programming is used to minimise the expected cost under variance constraints, the recalculated variances increase a little at the budget limit for each combination of misclassification and variable, compared to *ii*) above, but the variances are always at least as small as when the *SI*, *SI* design is used.

iv) The minimising of a simple variance function for each combination of misclassification and variable gives some variances that are slightly larger than when the *SI*, *SI* design is used, and some variances that are as small as in *ii*).

When a multicharacter survey is planned, the convex programming method is probably the most flexible of the methods shown in this chapter. The cost is minimised subject to constraints on the variances and sample sizes but it is quite possible to recalculate the solutions at the point where the available budget C_0 is exhausted.

In table 8.5 the percentage relative gains, $100AV(\hat{t}_{dr})/V_*(\hat{t}_d)$, at the expected cost C_0=2,500 are shown. As one could expect, the largest gains are obtained when the *SI*, *STSI* design is used, combined with optimum allocation separately for each parameter. However, these gains cannot be realised simultaneously. Comparing the convex programming solution and the minimum variance solution, the average gain is larger when the variance function is used, although the difference is small.

Table 8.1 *Coefficient of reliability, R_d^2, optimum solutions and relative gains in cost or variance for each combination of misclassification cases, variables and domains, when the SI design is used in both phases.*

					Minimising AV		*Minimising EC*		C^*/C_0 or
Case	*Var*	*d*	R_d^2	$\nu(d)$	$n_a^0(d)$	$AV^0(d)$	$n_a^*(d)$	$C^*(d)$	AV^0/V_*
A)	y_1	1	0.431	0.257	407	21,831	331	2,031	81%
		2	0.454	0.245	423	34,438	335	1,979	79%
	y_2	1	0.558	0.199	502	18,599	347	1,730	69%
		2	0.727	0.137	668	22,132	340	1,272	51%
B)	y_1	1	0.530	0.211	479	19,355	345	1,801	72%
		2	0.535	0.208	483	31,091	346	1,787	71%
	y_2	1	0.482	0.232	443	20,574	340	1,914	77%
		2	0.681	0.153	615	24,455	346	1,405	56%

Table 8.2 *Optimum solutions and relative gains in cost or variance for each combination of misclassification cases, variables and domains when the SI design is used in the first-phase sample and the STSI design is used in the second-phase sample.*

					Minimising AV		*Minimising EC*		C^*/C_0 or
Case	*Var*	d	$\nu_1(d)$	$\nu_2(d)$	$n_a^0(d)$	$AV^0(d)$	$n_a^*(d)$	$C^*(d)$	AV^0/V_*
A)	y_1	1	0.394	0.141	450	17,826	299	1,658	66%
		2	0.266	0.233	424	34,306	335	1,971	79%
	y_2	1	0.295	0.124	539	16,108	323	1,498	60%
		2	0.203	0.085	713	19,406	318	1,115	45%
B)	y_1	1	0.325	0.115	528	15,874	312	1,476	59%
		2	0.194	0.215	484	31,030	345	1,783	71%
	y_2	1	0.339	0.149	474	17,988	318	1,673	67%
		2	0.224	0.099	653	21,671	325	1,245	50%

Table 8.3 *Optimal values of* n_a, ν_1 *and* ν_2 *for each combination of misclassification cases and variables when convex programming and variance functions are used, minimum solutions and recalculated solutions at the cost* C_0=2,500.

		Variance function			*Convex programming*					
Case	*Var*	n_a^0	ν_1	ν_2	n_a^*	ν_1	ν_2	C^*	C^*/C_0	n_a^+
A)	y_1	426	0.319	0.204	334	0.226	0.233	1,971	79%	424
	y_2	639	0.236	0.099	323	0.295	0.123	1,498	60%	539
B)	y_1	486	0.252	0.184	345	0.212	0.207	1,788	72%	482
	y_2	577	0.265	0.184	317	0.339	0.149	1,673	67%	474

Table 8.4 *Variances and relative gains in domains* 1 *and* 2 *for each combination of misclassification cases and variables when convex programming and variance functions are used, minimum solutions and recalculated solutions at the cost* C_0=2,500.

			Variance function		*Convex programming*		
Case	*Var*	*d*	$AV^0(d)$	$AV^0(d)/V_*$	$AV^*(d)$	$AV^+(d)$	AV^+/V_*
A)	y_1	1	19,038	71%	26,411	20,819	77%
		2	34,974	80%	43,520	34,305	79%
	y_2	1	16,241	60%	26,880	16,106	60%
		2	19,498	45%	33,314	19,962	46%
B)	y_1	1	17,532	65%	26,880	19,224	72%
		2	31,999	74%	43,520	31,125	72%
	y_2	1	18,159	68%	26,880	17,988	67%
		2	21,784	50%	33,460	22,391	51%

Table 8.5 *The percentage relative gains,* $100AV(\hat{t}_{dr})/V_*(d)$, *at the cost* C_0=2,500 *for different allocation methods and two-phase designs.*

Misclassification	**A)**				**B)**			
Variable	y_1		y_2		y_1		y_2	
Domain	1	2	1	2	1	2	1	2
SI, SI	81%	79%	69%	51%	72%	71%	77%	56%
SI, STSI (opt alloc)	66%	79%	60%	45%	59%	71%	67%	50%
SI, STSI (var function)	71%	80%	60%	45%	65%	74%	68%	50%
SI, STSI (convex pgm)	77%	79%	60%	46%	72%	72%	67%	51%

BIBLIOGRAPHY

Andersson, C. (1992) Estimation vid felklassificerade data. Licentiatavhandling, Statistiska Institutionen, Uppsala Universitet.

Andersson, C. & Nordberg, L. (1994) A method for variance estimation of non-linear functions of totals in surveys – theory and a software implementation. (To appear in) *Journal of Official Statistics.*

Armstrong, J. B. & Wu, C. F. J. (1992) A sample allocation method for two-phase survey designs. *Survey Methodology*, 18, 253–262.

Bethel, J. W. (1989) Sample allocation in multivariate surveys. *Survey Methodology*, 15, 47–57.

Bethlehem, J. G. (1988) Reduction of nonresponse bias through regression estimation. *Journal of Official Statistics*, 4, 251–260.

Biemer, P. P. & Forsman, G. (1992) On the quality of reinterview data with application to the current population survey. *JASA*, 87, 915–923.

Biemer, P. & Stokes, S.L. (1991) Approaches to the modeling of measurement errors. In: P. P. Biemer *et al.* (eds.) *Measurement Errors in Surveys*, New York: Wiley.

Bross, I. (1954) Misclassification in 2 x 2 tables. *Biometrics,* 10, 478–486.

Chen, T. T. (1979) Log-linear models for categorical data with misclassification and double sampling. *JASA*, 74, 481–488.

Chiacchierini, R. P. & Arnold, J. C. (1977) A two-sample test for independence in 2 x 2 contingency tables with both margins subject to misclassification. *JASA*, 72, 170–174.

Chromy, J. R. (1987) Design optimization with multiple objectives. *Proceedings of the Survey Research Section, American Statistical Association*, 194–199.

Chua, T. C. & Fuller, W. A. (1987) A model for multinomial response error applied to labor flows. *JASA,* 82, 46–51.

Cochran, W. G. (1968) Errors of measurement in statistics. *Technometrics*, 10, 637–666.

Cox, D. R. & Hinkley, D. V. (1974) *Theoretical Statistics*. London: Chapman and Hall.

Deming, W. E. (1977) An essay on screening or on two-phase sampling, applied to surveys of a community. *International Statistical Review*, 45, 29–37.

Ekholm, A. & Palmgren, J. (1987) Correction for misclassification using doubly sampled data. *Journal of Official Statistics*, 3, 419–429.

Fuller, W.A. (1986) *Measurement Error Models*. New York: Wiley.

Haitovsky, H. & Rapp, J. (1992) Conditional resampling for misclassified multinomial data with application to sampling inspection. *Technometrics*, 34, 473–483.

Hájek, J. (1971) Comment on a paper by D. Basu. In: Godambe, V. P. & Sprott, D. A. (eds.) *Foundations of Statistical Inference*. Toronto: Holt, Rinehart and Winston, p. 236.

Hansen, M. H., Hurwitz, W. N. & Bershad, M. A. (1961) Measurement errors in censuses and surveys. *Bulletin of ISI*, 38, 359–374.

Hansen, M. H., Hurwitz, W. N. & Madow, W. G. (1953) *Sample Survey Methods and Theory*, Vol I. New York: Wiley.

Hochberg, Y. (1977) On the use of double sampling schemes in analyzing categorical data with misclassification errors. *JASA*, 72, 914–921.

Koch, G. G., Freeman, D. H., Jr. & Freeman, J. L. (1975) Strategies in the multivariate analysis of data from complex surveys. *International Statistical Review*, 43, 59–78.

Mak, T. K. (1988) Estimating subgroup means with misclassification. *JRSS, Ser B*, 50, 83–92.

Mak, T. K. & Li, W. K. (1988) A new method for estimating subgroup means under misclassification. *Biometrika*, 75, 105–111.

Nathan, G. & Eliav, T. (1988) Comparison of measurement errors for telephone interviewing and home visits by misclassification models. *Journal of Official Statistics*, 4, 363–374.

Neyman, J. (1938) Contribution to the theory of sampling human populations. *JASA*, 33, 101–116.

Raj, D. (1968) *Sampling Theory*. New York: McGraw-Hill.

Rosén, B. (1987) Some optimality problems when estimating household data on the basis of a "primary" stratified sample of individuals. R&D Report U/STM 36, Statistics Sweden.

Särndal, C. E. & Swensson, B. (1987) A general view of estimation for two phases of selection with applications to two-phase sampling and nonresponse. *International Statistical Review*, 55, 279–294.

Särndal, C. E., Swensson, B. & Wretman, J. (1992) *Model Assisted Survey Sampling*. New York: Springer-Verlag.

Selén, J. (1986) Adjusting for errors in classification and measurement in the analysis of partly and purely categorical data. *JASA*, 81, 75–81.

— (1990) Estimation of errorprone multinomial response. *International Conference on Measurement Errors in Surveys, Tucson.*

Smith, P. J. (1989) Is two-phase sampling really better for estimating age composition? *JASA*, 84, 916–921.

Statistics Sweden (1992) Evalvering av FoB90, hushållsdata och bostadsdata. Be 43 SM 9201, Statistics Sweden.

Swensen, A. R. (1988) Estimating change in a proportion by combining measurements from a true and a fallible classifier. *Scandinavian Journal of Statistics*, 15, 139–145.

Swensson, B. (1973) Estimation in the presence of non-sampling errors, a double sampling scheme with difference and regression estimation. Forskningsprojektet FEL I UNDERSÖKNINGAR, Rapport nr 62, Stockholms Universitet.

Tenenbein, A. (1970) A double sampling scheme for estimating from binomial data with misclassifications. *JASA*, 65, 1351–1361.

— (1971) A double sampling scheme for estimating from binomial data with misclassifications: sample size determination. *Biometrics*, 27, 935–944.

— (1972) A double sampling scheme for estimating from misclassified multinomial data with applications to sampling inspection. *Technometrics*, 14, 187–202.

Wolter, K. M. (1985) *Introduction to Variance Estimation*. New York: Springer-Verlag.

Woodruff, R. S. (1971) A simple method for approximating the variance of a complicated estimate. *JASA*, 66, 411–414.

Estimation of a Function of Totals

Consider the problem of estimating a parameter τ in a fixed population U of size N. The parameter τ is a function of J totals $\mathbf{t}=(t_1, \ldots, t_j, \ldots, t_J)'$, that is

$$\tau=f(t_1, \ldots, t_j, \ldots, t_J)=f(\mathbf{t}), \tag{A.1}$$

where $t_j = \Sigma_U y_{jk}$ is the total of the variable y_j in population U and y_{jk} is the value of y_j for element k. The vector $\mathbf{t}$ contains J totals generated by arbitrary combinations of y-variables and/or domains.

There are several ways to estimate τ, (see for example Wolter 1985). *One* natural estimator of τ is to replace the different totals in $f(\mathbf{t})$ by their estimates $\hat{\mathbf{t}} = (\hat{t}_1, \ldots, \hat{t}_j, \ldots, \hat{t}_J)'$,

$$\hat{\tau} = f(\hat{t}_1, \ldots, \hat{t}_j, \ldots, \hat{t}_J) \tag{A.2}$$

Note that $\hat{\tau}$ is generally not an unbiased estimator for τ when $f(\mathbf{t})$ is non-linear, even when $\hat{\mathbf{t}}$ is an unbiased estimator for $\mathbf{t}$. It is however a consistent estimator for τ if $\hat{\mathbf{t}}$ is a consistent estimator of $\mathbf{t}$. The domain mean where both denominator and numerator are estimated from the sample is one simple example on a non-linear function of two totals.

Several methods have been suggested for the estimation of the design variance of $\hat{\tau}$, but only the Taylor linearization method will be described here. The success of the method depends to a large extent on the assumption that the sample size is so large that the higher order terms of the Taylor approximation can be neglected. In that case we may write

$$\hat{\tau} - \tau \approx \sum_{j=1}^{J} f_j'(\mathbf{t})(\hat{t}_j - t_j) \tag{A.3}$$

where $f_j'(\mathbf{t}) = \dfrac{\partial f(\mathbf{t})}{\partial t_j}$ is the partial derivative of f evaluated at the point $\mathbf{t}$.

The approximate variance of $\hat{\tau}$ is calculated by

$$AV(\hat{\tau}) \approx V\left(\sum_{j=1}^{J} f_j'(\mathbf{t})\hat{t}_j \right) = \sum_{i=1}^{J}\sum_{j=1}^{J} f_i'(\mathbf{t}) f_j'(\mathbf{t}) C(\hat{t}_i, \hat{t}_j) \tag{A.4}$$

where $C(\hat{t}_i, \hat{t}_j)$ is the covariance between $\hat{t}_i$ and $\hat{t}_j$.

The approximate variance (A.4) is estimated by replacing $\mathbf{t}$ and $C(\cdot,\cdot)$ in (A.4) by their sample estimates

$$\hat{V}(\hat{\tau}) = \sum_{i=1}^{J} \sum_{j=1}^{J} f_j'(\hat{\mathbf{t}}) f_j'(\hat{\mathbf{t}}) \hat{C}(\hat{t}_i, \hat{t}_j) \tag{A.5}$$

In general, $\hat{V}(\hat{\tau})$ will not be an unbiased estimator of either $AV(\hat{\tau})$ or $V(\hat{\tau})$. It is however a consistent estimator of $AV(\hat{\tau})$ provided that $\hat{t}_j$ and $\hat{C}(\hat{t}_i, \hat{t}_j)$ are consistent estimators of t_j and $C(\hat{t}_i, \hat{t}_j)$. For a comprehensive discussion of the properties of (A.5), see Wolter (1985).

The transformation

$$u_k = \sum_{j=1}^{J} f_j'(\mathbf{t}) y_{jk} \tag{A.6}$$

is often used for computational convenience. Then the total $t_u = \sum_U u_k$ is estimated by $\hat{t}_u$ and the variance $V(\hat{t}_u)$ is an approximation of $AV(\hat{\tau})$.

The vector $\mathbf{t}$ in (A.6) is replaced by sample-based estimates, resulting in the variable

$$\hat{u}_k = \sum_{j=1}^{J} f_j'(\hat{\mathbf{t}}) y_{jk} \tag{A.7}$$

The variance of the estimated total $\hat{t}_{\hat{u}} = \sum_s \hat{u}_k / \pi_k$ is used as an estimator of $V(\hat{\tau})$. This technique was suggested by Woodruff (1971).

A computer program that uses (A.2) and (A.7) for the estimation of an arbitrary rational function of totals and its variance is described in Andersson & Nordberg (1994). It does not require explicit expressions for the derivatives .

When the *m*GREG estimator is used in combination with a two-phase design, the transformations (A.6) and (A.7) must be done for the *y*-values and the residuals *e* separately:

$$\hat{u}_{yk} = \sum_{j=1}^{J} f_j'(\hat{\mathbf{t}}) y_{jk} \tag{A.8}$$

$$\hat{u}_{gek} = \sum_{j=1}^{J} f_j'(\hat{\mathbf{t}}) g_{jk} e_{jk} \tag{A.9}$$

where $e_{jk}=y_{jk}-\mathbf{x}'_k\mathbf{b}_j$, $k \in s$ and where g_{jk} is defined by (2.20) for each j or set equal to 1.

The variance of $\hat{\tau}$ is then estimated by

$$\hat{V}(\hat{\tau}) = \sum\sum_s \frac{\Delta_{akl}}{\pi_{kl}^*} \frac{\hat{u}_{yk}}{\pi_{ak}} \frac{\hat{u}_{yl}}{\pi_{al}} + \sum\sum_s \frac{\Delta_{kl|a}}{\pi_{kl|a}} \frac{\hat{u}_{gek}}{\pi_k^*} \frac{\hat{u}_{gel}}{\pi_l^*} \tag{A.10}$$

Appendix B

Approximate Expressions for $AV(\hat{t}_{dr})$ when Measurement Model m is Used

B1 *SI* Designs in the First and Second Phases

A first-phase *SI* sample s_a of size n_a is drawn. A second-phase sample s of size $n = \nu n_a$ is drawn according to the *SI* design, where $0<\nu\leq 1$ is an *a priori* fixed constant. In s_a the error-prone values (x_k, z_{dk}), $k\in s_a$ are observed. In s both (x_k, z_{dk}) and the true values (y_k, ζ_{dk}), $k\in s$ are observed. Define $x_{d'k}=x_k z_{d'k}$ and $y_{dk}=y_k\zeta_{dk}$.

Under the model ξ with $E_\xi(y_{dk}|\mathbf{x}_k) = \mathbf{x}'_k \boldsymbol{\beta}_d$ and variance structure $\sigma^2_{dk} = \sigma^2_d \mathbf{j}' \mathbf{x}_k$, an approximately unbiased estimator of $t_d = \Sigma_{U_d} y_k$ is

$$\hat{t}_{dr} = \sum_{d'=1}^{D} \hat{t}_{axd'} \frac{\hat{t}_{ydd'}}{\hat{t}_{xd'}} \tag{B.1}$$

with $$\hat{t}_{axd'} = \frac{N}{n_a}\Sigma_{s_a} x_{d'k} \ ; \ \hat{t}_{xd'} = \frac{N}{n}\Sigma_s x_{d'k} \ ; \ \hat{t}_{ydd'} = \frac{N}{n}\Sigma_s y_{dk} z_{d'k} \tag{B.2}$$

The approximate variance of $\hat{t}_{dr}$, $AV(\hat{t}_{dr})$ is

$$AV(\hat{t}_{dr}) = \frac{N^2}{n_a}\left(1-\frac{n_a}{N}\right)S^2_{yd} + \frac{N^2}{n_a}\left(\frac{1}{\nu}-1\right)E_{SI}E_{\mathbf{X}}\left[s^2_{Ed}\right] = V_1 + V_2 \tag{B.3}$$

where

$$s^2_{Ed} = \frac{1}{n_a-1}\Sigma_{s_a}\left(E_{dk}-\overline{E}_d\right)^2 \ ; \ E_{dk} = y_{dk} - \mathbf{x}'_k \mathbf{b}_{ad} \ ; \ \overline{E}_d = \frac{1}{n_a}\Sigma_{s_a} E_{dk} \tag{B.4}$$

Element d' in $\mathbf{b}_{ad}$ for a given s_a and given values $\mathbf{x}_k$ is calculated as

$$b_{add'} = \frac{\Sigma_{s_a} y_{dk} z_{d'k}}{\Sigma_{s_a} x_{d'k}} \tag{B.5}$$

The variance s^2_{Ed} for a given s_a and $\mathbf{x}_k$ may be expressed as

$$s^2_{Ed} = s^2_{ayd} + \sum_{d'}\left(b^2_{add'}s^2_{axd'} - 2b_{add'}s_{aydxd'}\right) + \sum_{d'\neq d''}\sum b_{add'}b_{add''}s_{axd'xd''} \tag{B.6}$$

where $$s_{ayd}^2 = \frac{1}{n_a - 1}\Sigma_{s_a}\left(y_{dk} - \bar{y}_{ad}\right)^2 \;\; ; \;\; \bar{y}_{ad} = \frac{1}{n_a}\Sigma_{s_a} y_{dk} \tag{B.7}$$

The expression $E_{\mathbf{X}}\left[s_{Ed}^2\right]$ is evaluated by using the measurement model *m* in section 1.5 and the Taylor expansion of s_{Ed}^2, then

$$E_{\mathbf{X}}\left[\Sigma_{s_a} y_{dk} z_{d'k}\right] \approx \frac{N}{n_a}\Sigma_{s_a} y_{dk}\theta_{d'k} \tag{B.8}$$

$$E_{\mathbf{X}}\left[\Sigma_{s_a} x_{d'k}\right] \approx \frac{N}{n_a}\Sigma_{s_a} \mu_{xk}^{(1)}\theta_{d'k} \tag{B.9}$$

$$E_{\mathbf{X}}\left[s_{axd'}^2\right] \approx \frac{1}{n_a - 1}\left\{\Sigma_{s_a}\left(\sigma_{xk}^2\theta_{d'k} + \left(\mu_{xk}^{(1)}\right)^2\theta_{d'k}\right) - \frac{1}{n_a}\left(\Sigma_{s_a}\mu_{xk}^{(1)}\theta_{d'k}\right)^2\right\} \tag{B.10}$$

$$E_{\mathbf{X}}\left[s_{aydxd'}\right] \approx \frac{1}{n_a - 1}\left\{\Sigma_{s_a} y_{dk}\mu_{xk}^{(1)}\theta_{d'k} - \frac{1}{n_a}\Sigma_{s_a} y_{dk}\,\Sigma_{s_a}\mu_{xk}^{(1)}\theta_{d'k}\right\} \tag{B.11}$$

and $$E_{\mathbf{X}}\left[s_{axd'xd''}\right] \approx -\frac{1}{n_a(n_a - 1)}\left(\Sigma_{s_a}\mu_{xk}^{(1)}\theta_{d'k}\right)\left(\Sigma_{s_a}\mu_{xk}^{(1)}\theta_{d''k}\right) \tag{B.12}$$

The expectation with respect to *SI* is easily obtained by approximating $n_a - 1 \approx n_a$, replacing n_a with *N* and Σ_{s_a} with Σ_U . The component V_2 in (B.3) is approximated by AV_2, where $E_{SI}E_{\mathbf{X}}\left[s_{Ed}^2\right]$ is obtained by using (B.8)-(B.12). The accuracy of the approximation is demonstrated by a simulation study.

The same data, with the same classification and measurement characteristics, as in chapter 5 were used in a study of the accuracy of the approximations (B.8)-(B.12). The subsampling fraction ν was 0.2 (n_a=500, n=100). When variable y_2 was used, the errors in y_k were normally distributed with expectation zero and standard deviation $0.1y_k$. For each combination of variables y_1 and y_2 and classification cases A) and B) the theoretical value of $V_1 + AV_2$ ("Theoretical") was calculated according to (B.8)-(B.12). These calculations were compared with the means of 1,000 samples for each component in the estimated variances $\hat{V}_1$, $A\hat{V}_2$ ("Simulated") of which were estimated according to (3.25), and with the variances of the estimated totals.
The results of the simulations show that the approximations work reasonably well in the studied cases. The theoretical and simulated values of V_1 should be equal since $\hat{V}_1$ is an unbiased estimator of V_1.

Table B1. *Results of a simulation study of the accuracy of the approximate expressions for the variance of the estimates of the total of variable* y_1.

Case	*Domain (d)*	*Value*	V_1	AV_2	V_1+AV_2	$S^2(\hat{t}_{dr})$
A)	1	*Theoretical*	6,049.2	15,300.0	21,349.2	
		Simulated	6,090.9	14,994.9	21,085.8	22,083.0
	2	*Theoretical*	9,794.0	23,760.0	33,554.0	
		Simulated	9,779.2	23,788.5	33,567.5	34,584.5
B)	1	*Theoretical*	6,049.2	12,644.9	18,694.1	
		Simulated	6,090.9	12,381.5	18,472.3	19,260.0
	2	*Theoretical*	9,794.0	20,220.0	30,014.0	
		Simulated	9,779.2	20,311.0	30,090.4	30,856.5

Table B2. *Results of a simulation study of the accuracy of the approximate expressions for the variance of the estimates of the total of variable* y_2.

Case	*Domain (d)*	*Value*	V_1	AV_2	V_1+AV_2	$S^2(\hat{t}_{dr})$
A)	1	*Theoretical*	6,047.8	11,880.0	17,927.8	
		Simulated	6,056.4	11,620.5	17,677.0	18,732.0
	2	*Theoretical*	9,799.8	11,880.0	21,671.8	
		Simulated	9,781.8	11,935.6	21,717.4	22,269.0
B)	1	*Theoretical*	6,047.8	13,920.0	19,967.8	
		Simulated	6,056.4	13,657.9	19,714.0	20,862.0
	2	*Theoretical*	9,799.8	13,900.0	23,691.8	
		Simulated	9,781.8	14,004.2	23,786.0	24,138.1

B2 *SI* Design in the First Phase and *STSI* Design in the Second Phase

In the first-phase sample s_a, D classes are identified by using the observed z_{dk}, $k \in s_a$. The identified classes define the predetermined stratification of every sample s_a before the subsampling in the second phase. Let $n_{ad'}$ be the size of $s_{ad'}$, and assume that a subsample $s_{d'}$ of size $n_{d'} = \nu_{d'} n_{ad'}$ is drawn under the *SI* design, where $0 < \nu_{d'} \leq 1$,

(d'=1, ..., D), are *a priori* fixed constants. Then, if $n_{ad'}=0$ with probability zero, the component V_2 in $AV(\hat{t}_{dr})$ is approximated by

$$V_2 \approx \frac{N^2}{n_a} \sum_{d'=1}^{D} \left(\frac{1}{\nu_{d'}} - 1 \right) E_{SI} E_{\mathbf{X}} \left[\frac{n_{ad'}}{n_a} s_{Edd'}^2 \right] \tag{B.13}$$

where

$$s_{Edd'}^2 = \frac{1}{n_{ad'} - 1} \Sigma_{s_{ad'}} \left(E_{dk} - \bar{E}_{dd'} \right)^2 \quad ; \quad \bar{E}_{dd'} = \frac{1}{n_{ad'}} \Sigma_{s_{ad'}} E_{dk} \tag{B.14}$$

The variance $s_{Edd'}^2$ for a given s_a and $\mathbf{x}_k$, may be expressed as

$$s_{Edd'}^2 = s_{aydd'}^2 + b_{add'}^2 s_{axd'}^2 - 2 b_{add'} s_{aydxd'} \tag{B.15}$$

The expected value $E_{SI}E_{\mathbf{X}}[\cdot]$ in B.13 is evaluated by using the measurement model m in section 1.5 and the Taylor expansion of $s_{Edd'}^2$. By approximating $n_{ad'}-1 \approx n_{ad'}$ and noting that $E_{SI}E_X[n_{ad'}] = n_a \Sigma_U \theta_{d'k} / N = n_a \pi_{d'}$ the following, approximate, expressions are obtained,

$$S_{ydd'}^2 = E_{SI} E_{\mathbf{X}} \left[\frac{n_{ad'}}{n_a} s_{aydd'}^2 \right] \approx \frac{1}{N} \Sigma_U y_{dk}^2 \theta_{d'k} - \frac{1}{\pi_{d'}} \left(\frac{1}{N} \Sigma_U y_{dk} \theta_{d'k} \right)^2 \tag{B.16}$$

$$S_{xd'}^2 = E_{SI} E_{\mathbf{X}} \left[\frac{n_{ad'}}{n_a} s_{axd'}^2 \right] \approx \frac{1}{N} \Sigma_U \left(\sigma_{xk}^2 \theta_{d'k} + \left(\mu_{xk}^{(1)} \right)^2 \theta_{d'k} \right) - \frac{1}{\pi_{d'}} \left(\frac{1}{N} \Sigma_U \mu_{xk}^{(1)} \theta_{d'k} \right)^2 \tag{B.17}$$

$$S_{ydxd'} = E_{SI} E_{\mathbf{X}} \left[\frac{n_{ad'}}{n_a} s_{aydxd'} \right] \approx \frac{1}{N} \Sigma_U y_{dk} \mu_{xk}^{(1)} \theta_{d'k} - \frac{1}{N^2 \pi_{d'}} \Sigma_U y_{dk} \theta_{d'k} \Sigma_U \mu_{xk}^{(1)} \theta_{d'k} \tag{B.18}$$

$$b_{Udd'} = E_{SI} E_{\mathbf{X}} \left[b_{add'} \right] \approx \frac{\Sigma_U y_{dk} \theta_{d'k}}{\Sigma_U \mu_{xk}^{(1)} \theta_{d'k}} \tag{B.19}$$

$$b_{Udd'}^2 \approx E_{SI} E_{\mathbf{X}} \left[b_{add'} \right]^2 \tag{B.20}$$